Continuous Manufacturing *for the* Modernization *of* Pharmaceutical Production

PROCEEDINGS OF A WORKSHOP

Joe Alper, *Rapporteur*

Board on Chemical Sciences and Technology

Division on Earth and Life Studies

The National Academies of
SCIENCES · ENGINEERING · MEDICINE

THE NATIONAL ACADEMIES PRESS
Washington, DC
www.nap.edu

THE NATIONAL ACADEMIES PRESS 500 Fifth Street, NW Washington, DC 20001

This activity was supported by Contract/Grant No. HHSP233201400020B between the National Academy of Sciences and the Department of Health and Human Services. Any opinions, findings, conclusions, or recommendations expressed in this publication do not necessarily reflect the views of any organization or agency that provided support for the project.

International Standard Book Number-13: 978-0-309-48781-8
International Standard Book Number-10: 0-309-48781-1
Digital Object Identifier: https://doi.org/10.17226/25340

Additional copies of this publication are available from the National Academies Press, 500 Fifth Street, NW, Keck 360, Washington, DC 20001; (800) 624-6242 or (202) 334-3313; http://www.nap.edu.

Printed in the United States of America

Suggested citation: National Academies of Sciences, Engineering, and Medicine. 2019. *Continuous Manufacturing for the Modernization of Pharmaceutical Production: Proceedings of a Workshop.* Washington, DC: The National Academies Press. doi: https://doi.org/10.17226/25340.

The National Academies of

SCIENCES • ENGINEERING • MEDICINE

The **National Academy of Sciences** was established in 1863 by an Act of Congress, signed by President Lincoln, as a private, nongovernmental institution to advise the nation on issues related to science and technology. Members are elected by their peers for outstanding contributions to research. Dr. Marcia McNutt is president.

The **National Academy of Engineering** was established in 1964 under the charter of the National Academy of Sciences to bring the practices of engineering to advising the nation. Members are elected by their peers for extraordinary contributions to engineering. Dr. C. D. Mote, Jr., is president.

The **National Academy of Medicine** (formerly the Institute of Medicine) was established in 1970 under the charter of the National Academy of Sciences to advise the nation on medical and health issues. Members are elected by their peers for distinguished contributions to medicine and health. Dr. Victor J. Dzau is president.

The three Academies work together as the **National Academies of Sciences, Engineering, and Medicine** to provide independent, objective analysis and advice to the nation and conduct other activities to solve complex problems and inform public policy decisions. The National Academies also encourage education and research, recognize outstanding contributions to knowledge, and increase public understanding in matters of science, engineering, and medicine.

Learn more about the National Academies of Sciences, Engineering, and Medicine at **www.nationalacademies.org**.

The National Academies of

SCIENCES • ENGINEERING • MEDICINE

Consensus Study Reports published by the National Academies of Sciences, Engineering, and Medicine document the evidence-based consensus on the study's statement of task by an authoring committee of experts. Reports typically include findings, conclusions, and recommendations based on information gathered by the committee and the committee's deliberations. Each report has been subjected to a rigorous and independent peer-review process and it represents the position of the National Academies on the statement of task.

Proceedings published by the National Academies of Sciences, Engineering, and Medicine chronicle the presentations and discussions at a workshop, symposium, or other event convened by the National Academies. The statements and opinions contained in proceedings are those of the participants and are not endorsed by other participants, the planning committee, or the National Academies.

For information about other products and activities of the National Academies, please visit www.nationalacademies.org/about/whatwedo.

PLANNING COMMITTEE FOR A WORKSHOP ON CONTINUOUS MANUFACTURING FOR THE MODERNIZATION OF PHARMACEUTICAL PRODUCTION

Members

GINTARAS V. REKLAITIS (*Chair*), Purdue University
LIUQUAN CHANG, Merck
CHARLES COONEY, Massachusetts Institute of Technology
EVA GEFROH, Just Biotherapeutics
KEN LEE, MedImmune
KEITH ROPER, University of Arkansas

Staff

ELIZABETH BOYLE, Program Officer
MARILEE SHELTON-DAVENPORT, Senior Program Officer
ANNA SBEREGAEVA, Associate Program Officer
JARRETT NGUYEN, Senior Program Assistant

Consultant

JOE ALPER, Rapporteur

Acknowledgment of Reviewers

This Proceedings of a Workshop was reviewed in draft form by individuals chosen for their diverse perspectives and technical expertise. The purpose of this independent review is to provide candid and critical comments that will assist the National Academies of Sciences, Engineering, and Medicine in making each published proceedings as sound as possible and to ensure that it meets the institutional standards for quality, objectivity, evidence, and responsiveness to the charge. The review comments and draft manuscript remain confidential to protect the integrity of the process.

We thank the following individuals for their review of this proceedings:

Rubin Carbonell, North Carolina State University
Eva Gefroh, Just Biotherapeutics
Mike Ladisch, Purdue University

Although the reviewers listed above provided many constructive comments and suggestions, they were not asked to endorse the content of the proceedings nor did they see the final draft before its release. The review of this proceedings was overseen by **Stephen W. Drew**, Drew Solutions LLC. He was responsible for making certain that an independent examination of this proceedings was carried out in accordance with standards of the National Academies and that all review comments were carefully considered. Responsibility for the final content rests entirely with the rapporteur and the National Academies.

Contents

APPENDIXES

CONTINUOUS MANUFACTURING FOR THE MODERNIZATION OF PHARMACEUTICAL PRODUCTION: PROCEEDINGS OF A WORKSHOP

INTRODUCTION

On July 30-31, 2018, the National Academies of Sciences, Engineering, and Medicine held a workshop titled Continuous Manufacturing for the Modernization of Pharmaceutical Production. The workshop was convened under the auspices of the Board on Chemical Sciences and Technology and was sponsored by the U.S. Food and Drug Administration (FDA) and the Biomedical Advanced Research and Development Authority (BARDA). This workshop[1] discussed the business and regulatory concerns associated with adopting continuous manufacturing techniques to produce biologics such as enzymes, monoclonal antibodies, and vaccines (see Box 1, Statement of Task). The Statement of Task mentioned discussing upstream challenges of small molecules, however the planning committee determined that the discussions would be richer if focused solely on biologics. The workshop also discussed specific challenges for integration across the manufacturing system, including upstream and downstream processes, analytical techniques, and drug product development. The workshop addressed these challenges broadly across the biologics domain but focused particularly on drug categories of greatest FDA and industrial interest such as monoclonal antibodies and vaccines. The summary below describes the individual talks given at the workshop and tabular summaries of points made across themes are provided in Appendix C.

Public–Private Partnerships to Help Realize the Promise of Continuous Manufacturing

Kelvin Lee from the University of Delaware and the Director of the National Institute of Innovation in Manufacturing Biopharmaceuticals (NIIMBL) noted that continuous manufacturing is foundational to the production of iron, steel, paper, oil, cement, glass, synthetic fibers, electricity, clean water, and cars. For the same reason, it holds promise for the production of pharmaceuticals, which includes a lower cost of production with increased output to meet growing demand; the potential for higher and more reproducible quality; and superior

[1]This proceedings was prepared by the workshop rapporteur as a factual summary of what occurred at the workshop. The planning committee's role was limited to planning and convening the workshop. The views contained in the proceedings are those of individual workshop participants and do not necessarily represent the views of all workshop participants, the planning committee, or the National Academies of Sciences, Engineering, and Medicine.

productivity that uses equipment and space more efficiently. He added that while the demand for biopharmaceuticals is relatively small compared to the products listed above, the cost of manufacturing therapeutic proteins is high, and industry, patients, and developing nations would benefit from lowering that cost.

Mature processes—the production of iron or glass, for example—can implement continuous approaches more quickly because the chemical reactions are well-defined, the critical quality attributes (CQAs) are well characterized, and approaches to controlling the manufacturing process have been thoroughly tested. Continuous manufacturing processes are starting to emerge for small molecule drugs for the same reasons. The challenge with large molecule drugs, such as monoclonal antibodies, is that the biochemical process by which cells produce antibodies is not as well defined, and though there is a reasonable understanding of CQAs, the ability to control the production process is limited. These challenges are even more substantial for some biologics such as vaccines, cell-based therapies, and gene therapies. While there may not be business drivers in the near-term to explore producing these therapeutics via continuous manufacturing, continuous upstream processing of some biologics has become an accepted part of their production.

The challenges to adopting continuous manufacturing methods today include technological issues, quality and regulatory concerns, economic shortcomings, and perceived and actual risks. Regarding risk, Lee said that the robustness of continuous manufacturing methods as applied to therapeutic proteins is not well characterized, largely because of a lack of analytic and process control methods. Logistical concerns, such as the need for custom equipment and demands on the supply chain, can also increase the risk of transitioning from a bulk to a continuous manufacturing process. Given the pharmaceutical industry's aversion to risk, related to the large percentage of revenues reinvested in research and development, the high rate of failure at every stage of product development, and the years needed to bring a product to the market, industry's cautious approach to continuous manufacturing should not be surprising (Jacoby et al., 2015).

Going forward, while many regulatory agencies have expressed strong support for continuous manufacturing, adopting continuous processes for purifying an already marketed biologic product would mean going first, which is not the preference of most pharmaceutical companies. In addition, risk assessment likely overrides any cost-benefit calculation given that the revenue from a biologic is many multiples of the cost of goods. Industry's hesitancy is also related to its unwillingness to confront regulatory uncertainties and necessary changes to established quality systems associated with batch production capacity. Finally, said Lee, continuous manufacturing requires a high level of process understanding and control and highly robust and reliable methods for at-line process analytics and real-time monitoring that are still being developed. His solution to lessening the risk of developing the needed technologies and answering regulatory questions is to do so in consortia and public–private partnerships.

BOX 1 Statement of Task

An ad hoc committee will develop and conduct an interactive, multidisciplinary, multisector public workshop that focuses on the status of, and research opportunities for, continuous manufacturing in the pharmaceutical industry. The workshop will address biologics and upstream challenges of small molecules, with a focus on synergies in manufacturing the two kinds of products. The 2-day workshop will

1. Describe the current state of research and development for continuous manufacturing of pharmaceuticals, with an emphasis on (1) continuous bioprocessing (including control strategy) and (2) integrated upstream small molecule synthesis and purification (including control strategies such as process analytical technology and real-time release testing strategies). Included will be a discussion of technical limitations and challenges that may prevent adoption of continuous manufacturing in the production of both small molecule and biological products.
2. Discuss innovative solutions and creative ideas for addressing these limitations and challenges and highlight key areas that might benefit from technology investment or mitigation strategies to encourage and support broad adoption and rapid implementation of continuous manufacturing in the commercial sector.
3. Consider technical and programmatic lessons learned from continuous manufacturing of small molecule pharmaceuticals that may be applicable to biologics as well as newly identified challenges associated with biologics' continuous processing. Discuss opportunities for collaboration across scientific/technical disciplines (nexus between biology/biochemistry and bio/chemical engineering), across sectors (academia/public versus industry), including federal entities as appropriate, to accelerate the development and adoption of continuous manufacturing for biologic drug products.
4. Consider business challenges and regulatory hurdles that may impact the ability or the decision for industry to adopt continuous manufacturing processes.

Workshop participants will include representatives from academia, the pharmaceutical industry, U.S. government agencies, and professional organizations. The committee will plan and organize the workshop, select and invite speakers and discussants, and moderate the discussions. A workshop proceedings document will be prepared by a rapporteur to describe workshop presentations, workshop discussions, and common themes that emerged in the context of the objectives. Findings, conclusions, and recommendations will not be included in the proceedings.

NIIMBL,[2] with $70 million in funding from the National Institute of Standards and Technology (NIST) and more than $180 million in commitments from its member partners, is an example of such a public-private partnership. Part of NIIMBL's mission, explained Lee, is to accelerate innovation in biopharmaceutical manufacturing and support the development of standards that enable more efficient and rapid manufacturing capabilities by bringing together partners from industry, academia, and regulatory agencies to solve common problems. NIIMBL is part of the Manufacturing USA Network, which formed to address market failure of insufficient industry research and development in the so-called advanced manufacturing "valley of death" between when technology is produced in the laboratory and when it is produced in a representative manufacturing environment, corresponding to Manufacturing Readiness Levels 4 through 7. By addressing this "missing middle," the network aims to reduce risk and accelerate the adoption of new manufacturing technologies and to simultaneously address any regulatory issues that may be associated with those new technologies.

The pharmaceutical industry, argued Lee, requires further industrialization in manufacturing technology if it is to meet variable market needs for lot size and demand, address speed to market, improve flexibility, and enable patient access to emerging therapies through the development of new automated, small-scale manufacturing platforms that are integrated with robust product and process measurement capabilities. Other than NIIMBL, there has been no coordinated effort to ensure that biomanufacturing keeps pace with an increasingly diversified product pipeline, to support the biomanufacturing supply chain in the United States, and to develop the knowledge base needed to mitigate the risk of adopting new processing technology in a highly regulated industry.

The Future of Access to Medical Countermeasures

Rick Bright from BARDA explained that his agency is interested in continuous manufacturing as a means of rapidly producing medical countermeasures to evolving and increasing natural and human-made health security threats. Established in 2006, BARDA's mission is to build unique public–private partnerships to bridge the manufacturing "valley of death" that Lee described. Congress has empowered BARDA with flexible and nimble authorities to work with industry in ways that were challenging for other federal entities. BARDA, for example, has the authority to fund projects for multiple years to create a strong commitment to its industry partners to address the most difficult challenges. It can also make direct hires of industry experts who can work hand in hand with its more than 200 industry partners on development and production of medical countermeasures. Evidence that this model works includes the 38 BARDA-supported FDA licensure and approvals for 35 different medical countermeas-

[2]See https://niimbl.force.com/s (accessed September 6, 2018).

ures, the addition of 14 products to the Strategic National Stockpile including 8 new FDA licensures, the 10-fold expansion of domestic influenza vaccine production capacity, and accelerated antibacterial product development to address critical vulnerabilities raised by antimicrobial resistance.

At the end of the day, said Bright, BARDA and its partners are addressing patient access to medical countermeasures in a timely manner to protect the health of Americans when a threat occurs. These efforts include tackling manufacturing challenges and working with FDA on regulatory issues to create a mechanism to allow the nation to mount a rapid response to any threat that comes its way. They also include strengthening the U.S. manufacturing base to address the fact that commercial manufacturing, for many reasons, has moved offshore. BARDA's ultimate goals are to be proactive, create a distributed manufacturing and delivery system, eliminate drug shortages, and limit the need to manufacture and stockpile medical countermeasures.

Bright noted that FDA has been a leader in the initiative to create the desired state in which every American has access to medical countermeasures when needed. Toward that end, BARDA and FDA established a continuous manufacturing partnership in 2015 to help drive this field. FDA is focused on developing the regulatory science to address operational and technical challenges, while BARDA focuses on evaluating continuous manufacturing processes, improving process efficiency, and ensuring the sustainability of medical countermeasure manufacturing. Ensuring drug access for all, said Bright, requires a continued strong partnership with FDA to address regulatory challenges as new technologies emerge. He encouraged everyone to work with FDA proactively.

Through the multi-agency National Biopharmaceutical Manufacturing Partnership, BARDA and the U.S. Department of Defense (DoD), with industry and academic partners, have built four new manufacturing facilities, each with the capacity to produce drugs and vaccines. Capacity is one thing, said Bright, but response capability is a different matter, so BARDA and DoD are in the process of building a unified, U.S. whole-of-government pharmaceutical manufacturing capability that would manufacture and have the ability to support the development of new drugs and vaccines, but also serve as a rapid turn-on switch that would function as surge capacity for the nation.

BARDA's newest initiative, empowered by the 21st Century Cures Act, is its Division of Research, Innovation, and Ventures (DRIVe), which has the mission of accelerating the research, development, and availability of transformative countermeasures. BARDA is using DRIVe to transform the way government works with industry to make it easier—and faster—to work with the federal government. For example, DRIVe intends to approve meritorious proposals within 30 days of receiving the proposal. More importantly, said Bright, DRIVe will focus on everything it can to increase timely access to medical countermeasures, including earlier detection of threats and earlier notification of individuals who have been exposed to a biological threat so that those who are exposed to an agent, such as pandemic influenza, can perhaps get treatment before symptoms appear and the infectious agent is spread from person to person.

Toward that end, DRIVe aims to create a "pharmacy on demand" that makes medicines readily available to everyone through emerging technologies, perhaps via a delivery service or a booth that dispenses medications via a smartphone eScript app.

FDA's Interest in Continuous Manufacturing

Janet Woodcock from FDA noted that biopharmaceutical manufacturing is an undervalued component of the biopharmaceutical industry and the nation's drug supply. As the nation deals with shortages of some drugs and issues with storage capacity, lack of surge capacity, and natural disasters such as hurricanes, the health care sector is concerned over its inability to care for its patients because the manufacturing sector has not been able to respond to these challenging circumstances. In addition, the global supply chain that helps reduce the cost of production has also increased the vulnerability of the nation's drug supplies.

As a result of these and other factors, FDA has long been interested in furthering the science of pharmaceutical manufacturing, starting in the small molecule space. In the early 2000s, FDA started its Pharmaceutical Manufacturing for the 21st Century initiative to bring the challenges of drug manufacturing to light. One response to this initiative from industry was that FDA would be resistant to changing manufacturing technologies, and Woodcock admitted there was some truth to that sentiment. This prompted FDA to adopt the role of advocate for change rather than be an obstacle to change. FDA began encouraging companies to adopt on-line monitoring, which required equipment manufacturers to develop new instrumentation. It also worked to harmonize international regulations of small molecule manufacturing to address barriers to continuous manufacturing.

Continuous manufacturing of both small and large molecules, said Woodcock, can dramatically shorten the time that it takes to scale up manufacturing for newly approved drugs. At a time when FDA is approving drugs designated as breakthrough products after an abbreviated clinical trial process, companies can have less time to develop commercial scale processes, which puts a premium on approaches that can scale quickly. FDA is now overseeing the development of continuous processes in the small molecule space, as well as model-based control strategies that would lead to real-time release of products, and it expects a handful of regulatory submissions over the coming year. The agency is also overseeing work being done on the continuous synthesis of biopolymers such as therapeutic DNA and RNA molecules. In summary, continuous manufacturing is gradually seeping into the small molecule space as commercial opportunities emerge, and the agency is prepared to handle and approve those applications.

In the biologics space, continuous manufacturing is more advanced at the upstream end of the process, and progress going forward requires integration with continuous downstream processing. Here, Woodcock noted, the barriers to adoption are more about the business case than technical challenges, though there are technical issues to address. The value proposition for business includes reducing the number of steps involved in manufacturing and the need for human handling during intermediate stages of production, improved safety, shortened processing time, smaller equipment and facility needs, more flexible operations, reduced capital expenditures, decreased environmental footprint, feasibility of manufacturing small batches economically, on-line monitoring for better quality control, and real-time release of final product.

One challenge according to Woodcock is to develop advanced control strategies and incorporate real-time data strategies for CQAs. Others are to integrate downstream unit operations in an effective manner that satisfies purity requirements and for the community to agree on real-time release methods. The real barriers, though, are questions that industry has about return on investment in continuous manufacturing and the established fixed infrastructure. Regulatory un-certainty is another barrier even though FDA is actively promoting this transition. It was suggested by an audience member that FDA could hold webinars with international regulators and industry to clarify where other nations stand on these technologies.

FDA's current approach to push this initiative, said Woodcock, is to collaborate with NIIMBL, BARDA, NIST, the Defense Advanced Research Projects Agency (DARPA), academia, and industry. When the initiative started, there was little academic work in this field, but FDA has been able to nurture academic research through a grant program on continuous manufacturing. Internally, FDA's emerging technology team has been working inside the agency to educate the regulators. In terms of next steps, FDA hopes that this workshop will move the field forward, and Woodcock stated that the agency is ready to work with any group that is developing continuous bioprocessing, even before the process is fully finished. She suspects that it is possible in the biologics space that some of the special concerns—anti-terrorism concerns, outbreaks, and shortages—will stimulate adoption of continuous bioprocessing approaches that will then diffuse into more standard bioprocessing.

Twenty years from now, biological product manufacturing is not likely to look like it does today. The field will advance, said Woodcock, and the outlines of where the field should go are already visible. The question is how fast will this evolution occur, and FDA believes that by working together, collaborating, and taking incremental steps, that future can occur faster.

BUSINESS CASE FOR CONTINUOUS MANUFACTURING

Transforming Biopharmaceutical Production Through the Deployment of Next-Generation Biomanufacturing

Art Hewig from Amgen said that biomanufacturing has been evolutionary when compared to other industries, and that a changing business landscape requires agility, flexibility, modularity, and dematerialization; or reducing the size of biomanufacturing networks. Continuous manufacturing can support this transition. He noted that the business case for adopting continuous manufacturing depends on the specific company and how that company operates. Making the business case, he said, "begins with understanding your business and your portfolio mix as it exists and as you see it coming, and where your manufacturing network is at, and what you think your overall plan for the future is." Intensification, or shrinking the manufacturing footprint, is an important aspect of a business case, but Hewig cautioned to make sure that the footprint of a redesigned continuous manufacturing process actually takes up a smaller footprint when all is said and done.

The biopharmaceutical landscape itself is changing in terms of its focus on patient needs, its move to flexible drug discovery and development, and its expanding global presence. As a result, product mix is becoming more heterogeneous with the development of more targeted products beyond monoclonal antibodies, demand uncertainty has increased, and the volume of product needed is dropping. Overall, this requires balancing the use of existing facilities and the investment made in those facilities against the addition of new capabilities to lower costs and increase flexibility and speed.

Hewig said that productivity improvements in fed-batch manufacturing processes have plateaued after three decades of commercial production of therapeutic proteins. As a result, companies have been turning to new technologies such as perfusion processes and high cell density processes to start shrinking the footprint from a productivity standpoint. Today, a highly productive 2,000 liter perfusion system can match the productivity of a 15,000 liter fed-batch bioreactor, which at that size creates the possibility of having single-use, disposable bioreactors. He noted that single use solutions also include mixing and chromatography systems, creating the opportunity to update the biomanufacturing paradigm to include off-the-shelf, deployable, and scalable technologies.

Amgen, he explained, started its efforts to revamp its manufacturing processes in 2010, and it expects this effort to take years to complete given the need to develop new processes and technologies and have them approved by regulatory authorities. This effort is paying off, though, and the company's next generation manufacturing facility in Singapore, which is 80 percent smaller than its older facilities, was approved for commercial manufacturing. The company is building a duplicate facility in Rhode Island.

The next evolution of biomanufacturing, Hewig predicted, will involve additional process intensification and integration that will result in further shrinking of the manufacturing footprint. This will allow companies to take a modular approach to scale out their production capacity to meet increasing demand. It will also enable companies to work at a commercial production scale earlier in the development process and avoid the uncertainties of moving from a pilot scale to a commercial scale reactor.

Turning to the business case, shrinking the manufacturing footprint results in a significant reduction in capital investment and the time needed to deploy a new process. It also enables miniaturization and intensification of process workflows. At the same time, the cost structure shifts from fixed to variable, and it allows for better targeting investment based on market demand and product mix. Technology transfer becomes less risky with the ability to scale out instead of scale up. He estimated that continuous manufacturing could reduce capital expenditures from $1 billion per facility to hundreds of millions to eventually tens of millions and the time for construction from years to months. What this allows from a business viewpoint is to delay deploying a new facility in the face of the uncertainty associated with introducing a new product to the market. This is a huge advantage to the current situation of having to decide to invest billions of dollars on building production capacity based on an assumption that a new drug will be a blockbuster long before it has even finished clinical trials. In his view, a hybrid network of conventional and flexible plants will create a responsive and efficient supply chain for therapeutic proteins.

In the end, the business case for continuous manufacturing is not a simple yes or no answer, said Hewig, and how it provides the greatest value will depend on the specific needs of each company (Pollock et al., 2017; Walther et al., 2015). Important considerations include a company's product portfolio, market demands, the existing manufacturing network, and the manufacturing business model. Continuous manufacturing does offer several opportunities, including the possibility of identifying a "best" approach to flexible mass output and developing improved single-use systems. Challenges include the need to balance labor costs and automation and integrate the different components of a continuous process, develop new control strategies, and understand the costs of analytical testing associated with continuous manufacturing.

Continuous Manufacturing for Large Molecule Drugs

Mauricio Futran from Janssen Pharmaceuticals reiterated that there is no single business case for continuous manufacturing of large molecules. Asset utilization, process flexibility, quality efficiency gains, and time to market are all possible drivers for deciding to use continuous manufacturing. In addition, the ability to monitor product quality during production, rather than at the end of a batch process, can improve resource and process efficiency. Futran added that a company should not implement a continuous process for the sake of having a continuous process—what problems is the company trying to solve? The hurdle to implementing continuous processes is easier with known unit operations transformed from batch to continuous, rather than starting from scratch.

From his perspective as a chemical engineer, Futran said the biggest benefit of continuous manufacturing is that all operations can be run simultaneously in a single room, enabling continuous monitoring of all production steps, something that is lost in batch production. A one-room situation for large molecule production could raise questions about contamination that could spread downstream, however, and the promised improvements in quality control efficiency and process robustness depends on developing in-line assays and methods for fully characterizing the entire process. Continuous manufacturing can enable more agile supply planning and response. It may also reduce the material needed for development and technology transfer.

Additional considerations such as the need to integrate continuous manufacturing with modern supply chain capacity management includes working with external contract manufacturers and dealing with starts and stops in the production process. From his experience with continuous manufacturing of small molecule drugs, startups and stops are difficult, take time, and have significant risk, and doing so with the living systems used to produce large molecules will be even more challenging. The ability to measure critical attributes for control is necessary, he added, and developing the appropriate methods will be challenging for a continuous bioreactor given that living systems are often not operating at a steady state.

Despite these challenges, continuous manufacturing of large molecules offers many benefits for the company, patients, and regulators. Enhanced quality control efficiency has the potential to reduce regulatory review timelines. Development times can be faster, improving speed to market, while continuous manufacturing of large molecules may provide benefits in terms of the ability to better manage demand fluctuations and the supply chain.

Continuous Processing Beyond Financials

Franqui Jimenez from Sanofi explained that there are two general approaches to continuous manufacturing: a fully integrated process that physically connects every step from media generation through purification, and a hybrid process that is integrated through the capture phase, with the product then feeding into subsequent batch purification processes. In both cases, the process is considered continuous if it comprises integrated, continuous unit operations with zero or minimal hold volume between them and balances mass and flow throughout the process (Konstantinov and Cooney, 2015).

When developing a continuous manufacturing process, it is important to consider that steady state conditions are expected to lead to consistent product quality, which could reduce or replace lot release testing. Development packages, however, require increased understanding of failure modes, which can be intensified in continuous manufacturing, as well as safe production stop points and the linkages between operations. Continuous processing, said Jimenez, produces large material volumes quickly, which increases the availability of material for characterization. On the other hand, moving to a continuous process may require more complex analytical capabilities with respect to instrumentation and technology. Additionally, it is expected that companies will need staff capable of responding to production issues associated with these advanced technologies, stated Jimenez.

Regarding operational considerations, Jimenez said that continuous processes require improved process controls, including the need for on-line, in-line, and at-line measurements, to maintain a steady state in the bioreactor. This is needed to lower the complexity of the process and reduce shop floor instructions. In addition, the manufacturing facility must be capable of running continuously for longer periods, requiring both more durable instrumentation and increased reliance on engineering controls to prevent contamination. In his opinion, the most prudent approach to developing a continuous process is to develop it in islands of continuous processes that can then be linked together as the understanding of those processes grows.

On a final note, Jimenez discussed the challenge that continuous manufacturing presents for regulation in terms of process verification. For example, the current paradigm requires running five batches to demonstrate process consistency, but this would become a demanding requirement for a continuous process, which may run for 60 days, with little gain. Jimenez described an example of what this new paradigm could look like: a process performance qualification run followed by two continuous process verification runs. Sanofi has been arguing that the increased process knowledge required for a continuous production scheme and the fact that process development occurs at the commercial scale offers the opportunity to create a new paradigm for approving a manufacturing process.

In a panel discussion following the three talks, moderated by Gintaras V. Reklaitis from Purdue University, the session speakers discussed that the challenges for moving existing approved products to continuous processes and establishing comparability between the processes is not trivial. The three speakers agreed that there are fewer drivers to implement fully continuous immobilized cell bioreactor compared to implementing a continuous perfusion bioreactor only, and that it is best to implement continuous processes in stages, not all at once. Hewig said that Amgen learned the benefits of single use facilities from the Singapore facility because modifications can be made in response to changes in the supply chains. In response to a question about what drove change at companies, Hewig said that Amgen did not want to build another stainless facility so they embraced single use. Jimenez said that at Sanofi, Konstantin had to champion and invest in the technology.

UPSTREAM PROCESSING

Fitting a Continuous Process into Existing Facilities

Daisie Ogawa from Boehringer Ingelheim reported on a collaborative project between Boehringer Ingelheim and Pfizer. For 3 years, the two companies have been partnering to research and develop radically cheaper and more rapidly produced clinical material. These speed and cost improvements allow research and development to explore additional clinical options and enable faster proof of concept with a clear path to commercialization. A major task for this project has been to develop an integrated skid based on single-use technology that includes a single-use bioreactor, elution stream chamber, and continuous viral inactivation up to the single mixer, which then feeds into a batch process viral reduction filter. The integrated pieces run continuously with no air gaps, she explained.

Perfusion cell culture, explained Ogawa, must still address two limitations relative to the fed-batch bioreactor in their facility: the unacceptably long run times to produce comparable amounts of material—30 to 90 days compared to 14 days—and the extremely large volumes of media required for a perfusion cell culture system. To address these limitations, she and her colleagues first worked to intensify the seed train. Traditional batch "N-1" seeding requires 7 to 10 days just to ramp up production, but using a perfusion N-1 stage produced a 10-fold higher N-stage seeding density, reducing the time to reach the production phase by approximately 5 days.

To shorten the production time further, Ogawa's team shifted to a non-steady state perfusion culture for which they do not control cell density. The challenge then became supplying adequate nutrients while maintaining a low perfusion rate and the solution was to use concentrated media and a diluent, feed them to the cells based on osmolality demands, and balance that with waste flushing. She noted that the logistical challenges of media preparation, storage, and transport are reduced substantially using concentrated feeds. For example, perfusing a 1,000-liter bioreactor over a 14-day process at two vessel volumes per day would take approximately 30,000 liters of media prepared at the usual concentration, but only 7,000 liters using concentrated feeds. This change reduces costs and allows for preparing the media onsite, and more importantly, generates high viable cell densities and 5-fold higher daily average productivity relative to a steady state system. Compared to a fed-batch process, the non-steady state bioreactor produced material equivalent to a 76 gram/liter fed-batch reactor, or greater than 10-fold higher productivity than a fed-batch system. This intensified process scaled successfully from a 2-liter reactor to a 100-liter reactor (see Figure 1).

Non-steady state perfusion using media concentrates produces more product per liter of media consumed than steady state perfusion. The resulting process looks more like a fed-batch system in that it operates over 14 days, and even though media usage is higher, it is still at a scale that it can be produced in the same 12-kiloliter facility with minimal capital expenditures. The challenge now, said Ogawa, is that the amount of material produced by this intensified process exceeds the purification train's maximum capacity. What Boehringer Ingelheim envisions is building a new facility using the integrated SKID with a 100-liter bioreactor to produce enough material for toxicology testing and a 1,000-liter bioreactor for commercial-scale production that would be capable of producing 1,200 kilograms of material per year.

In summary, Ogawa said that a non-steady state perfusion system using concentrated media feeds could fit into a commercial 12-kiloliter facility with little capital investment. In addition, this intensified, highly productive perfusion process allows one facility to produce multiple products, which is expected to translate into faster and less expensive production of biologics. Moreover, the integrated SKID platform can achieve commercial-scale production using 500 to 1,000-liter bioreactors, which is expected to lead to faster, more efficient, and less expensive production.

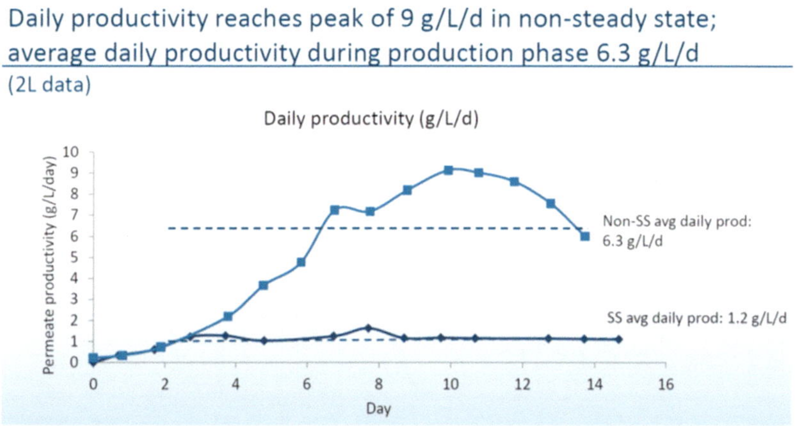

FIGURE 1 Daily productivity in a non-steady state production phase. NOTE: d refers to days, g refers to grams, L refers to liters, and SS refers to steady state. SOURCE: Ogawa, slide 15.

Intensification of a Multi-Product Perfusion Platform

Shawn Barrett from Sanofi explained that his company is moving from its legacy microcarrier Chinese Hamster Ovary (CHO) culture system with a batch downstream processing system to a single-use system with a suspension CHO culture and continuous capture to reduce cost of goods and process footprint and increase robustness and flexibility. Going forward, Barrett's upstream technology development team is working on perfusion intensification beyond integrated continuous biomanufacturing. The aim is to increase volumetric productivity, leverage in-house medium formulations, and reduce cell-specific perfusion rate to minimize media flow through the system, while maintaining desirable product quality attributes, process robustness over extended run times, and consistent cell densities in a system that can be scaled to support intensified processes. The technology development focus, he added, is on monoclonal antibodies and monoclonal antibody-like products that are in the company's early pipeline with the intent of minimizing or eliminating the need for process optimization in the early stage of development. The idea, explained Barrett, is to take a new cell line, put it in a perfusion bioreactor, and then go right to scale up.

The first version of Sanofi's integrated continuous biomanufacturing perfusion system (see Figure 2), using a new cell line and chemically defined medium, yielded a 100-fold improvement in productivity over the legacy process. Further development has produced as much as a 5-fold improvement at the 10-liter scale with other biologics. There were some issues when using different cell lines, resulting in unexpected growth arrest, lower cell viability, and a 40 to 60 percent decline in productivity. The latest platform version uses a new

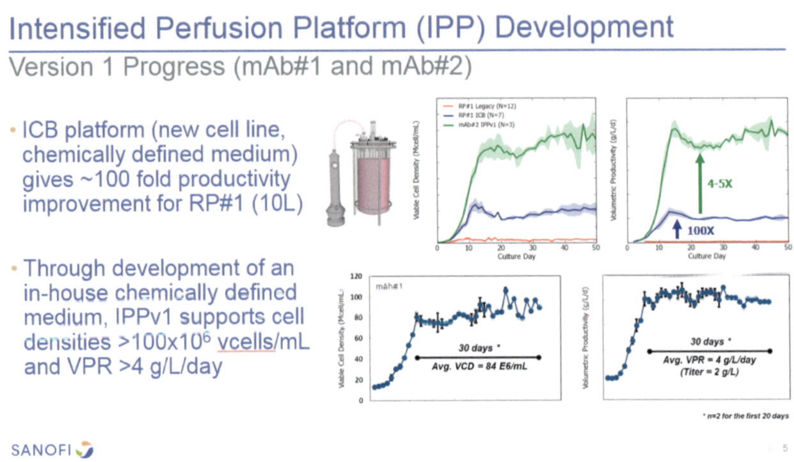

FIGURE 2 Sanofi's intensified perfusion platform. NOTE: ICB refers to integrated continuous biomanufacturing, mAb refers to monoclonal antibody, RP refers to reliable plant, VPR refers to high volumetric productivity. SOURCE: Barrett, slide 5.

perfusion supplement and additional concentrates as needed. He noted that all clones used in the intensified perfusion platform have been generated from a fed-batch clone screening workflow that the company would prefer not to change. The new version has now demonstrated with eight different clones that it can produce perfusion titers anywhere from half to equivalent to those obtained with a fed-batch system, with a specific productivity boost of 30 to 60 percent. Barrett described their current focus toward confirming scale-up readiness in 500-1,000 L single-use bioreactors (SUBs), and that one SUB vendor did not have enough gas mass transfer capacity and another was not a preferred vendor, so they have been working with their preferred vendor to modify the oxygen transfer rates to be more supportive of the cell densities they want to achieve.

Modeling the cost of goods for a generic monoclonal antibody produced using intensified perfusion integrated with continuous capture showed that a 2,000-liter perfusion system had the potential to produce a dramatic reduction in cost of goods. He noted that with increasing plant capacity, a 2.5-fold increase in titer could decrease cost of goods by approximately 50 percent.

Drivers, Challenges, and Implementation Solutions for Continuous Perfusion Manufacturing

Eva Gefroh from Just Biotherapeutics explained that her company's goal is to design and apply innovative technologies to expand global access to breakthrough biotherapeutics dramatically by reducing the cost of these products by at least a factor of 10. The company aspires to develop technology to reduce the time to an investigational new drug application for a monoclonal antibody to less than 10 months, produce more than 10 new molecules per year, and reduce the cost of goods to under $10 per gram, which has proven to be difficult. It also wants to enable smaller, more reconfigurable facilities that can be built at lower cost and on a much faster timeline.

To meet that goal, Just Biotherapeutics is designing a semi-continuous process with continuous capture chromatography and batch downstream processing for final purification and virus removal, similar to the approach Barrett discussed. The idea is to deploy this system in a small footprint facility containing modular, reconfigurable cleanroom pods that can be built and validated within 18 months. Modeling studies suggest that a high-productivity perfusion process can produce upward of 1,000 kilograms of product per year, with increasing volumetric productivity playing a large role in driving down the cost of goods manufactured. Flexible manufacturing, said Gefroh, is fundamental to managing demand uncertainty and driving toward a low cost of goods. It also enables a startup company such as hers to build a relatively low cost facility—she estimated the cost to be $60 million to $80 million—that can be modified and expanded as needed.

The activities that she and her colleagues have been undertaking to develop the company's continuous perfusion process platform include attaining low cell-specific perfusion rates, maintaining high cell density at a steady state through media development and cell bleed, developing the perfusion filter, and addressing the sensitivity of cells to shear conditions and pluronic concentration. They are also addressing the challenges that come with scale up from a 2-liter bioreactor to the 500-liter scale, including carbon dioxide stripping, vent filter sizing, and shear in the perfusion loop, oxygen availability, and agitation rate.

Handling the complexity of media usage is also a big development area. For example, at a 1,000-liter scale, media and buffer preparation would have to be scheduled in separate pods at close to full capacity. To reduce this demand, the company is developing proprietary, in-house produced media, working to reduce perfusion rates, and preparing media concentrates that can be stored at room temperature to reduce the high cost of refrigerated storage. Gefroh explained that a concentrated, stable at room temperature media would create the potential for outsourcing liquid media production to an outside vendor, which would reduce solution preparation areas and associated labor costs. Challenges in creating media concentrates include limited solubility at high concentrations, filterability, and the number of feed solutions required to have the desired solubility. Once those issues are solved, strategies for delivering the concentrated media to the bioreactor need to be worked out to ensure a consistent and accurate feed to the bioreactor. So far, results with a continuous perfusion process using concentrated media at the 500-liter scale have shown this approach to be feasible with respect to productivity over 21 days.

DOWNSTREAM PROCESSING

Practical Considerations for Adoption

Lindsay Arnold from MedImmune discussed her company's efforts to develop downstream processing for monoclonal antibody production on the assumption that it can operate regardless of the upstream system that feeds into it. She noted that the continuous downstream process—comprising multi-column chromatography for continuous Protein A capture, low pH virus inactivation in a packed column, a filter train, multi-column chromatography for continuous polishing, and single-pass buffer exchange and final concentration—has been run with continuous upstream manufacturing in a fully integrated system for 2 weeks.

The pilot downstream facility, she explained, uses a 50-liter perfusion bioreactor and can handle up to a 200-liter perfusion bioreactor, and it processes 200 to 300 liters of fed-batch material over 2 to 3 days. Each of the five components of the downstream system are currently automated individually, but the goal is to develop a fully integrated control system for all five units. Arnold explained that the primary capture stage is agnostic to the equipment in the upstream supply chain, but the company has developed its own approach to low pH virus inactivation using in-line titration followed by a static mixer. Residence time, she said, is achieved in a packed agarose column. The company is taking a plug-and-play approach for the filter train and wants to work with a vendor to supply the entire filtration-anion exchange membrane-virus filtration combination.

One of MedImmune's challenges in moving this system from the pilot scale to commercial production is that the company and its parent AstraZeneca currently have excess production capacity, which means that any changes become cost prohibitive given the existing installed capital base. Low utilization, said Arnold, does not push innovation. The strategy, then, is to take a modular, one-unit-at-a-time approach and introduce continuous processing in places where batch processing comes up short in terms of productivity and efficiency, thereby gaining experience and building a knowledge base for when the company needs to expand its production capabilities and build a new facility.

Given the opportunity for a large cost-of-goods savings in the initial capture step, Arnold and her team have been evaluating scale up opportunities for this unit. Going forward, she has been focusing on identifying other places where continuous bioprocessing can increase productivity and cost of goods, offer opportunities for enhancing process control, and where a particular molecule or therapeutic modality comes with specific processing needs, such as with labile molecules, cell and gene therapies, and non-monoclonal antibody modalities that would drive adoption of single-use systems.

In summary, Medimmune is not likely to adopt fully continuous biologics manufacturing in a 5-year period, but is likely to adopt it in a step-wise fashion with multi-column chromatography as a first step. The business cases, said

Arnold, will vary based on the particular sets of drivers that are important to different companies, such as safety, cash flow, capital reduction, space constraints, and capital improvement, and will vary between clinical and commercial settings. "The best we can do right now is prepare an almanac of technologies so that when there is a crisis, we can be there," she said.

What Can Be Learned from the Chemical Industry

Andrew Zydney from The Pennsylvania State University reiterated Lee's statement in the day's first presentation that many industries have converted from batch to continuous processing. The chemical industry, for example, made this transition nearly one century ago, which provided significant increases in productivity, reductions in pollution, improved product quality, and enhanced process safety. He explained that the chemical industry used multiple strategies to develop continuous processes, many of which are based on countercurrent staging. Countercurrent staging, he added, can also be used to develop continuous bioprocesses. Examples include diafiltration for buffer exchange and formulation, purification using continuous chromatography, and capture via continuous precipitation.

Conventional diafiltration, which is used to replace buffer from an upstream chromatography step with another buffer, is typically the final step in the formulation of a drug substance. As practiced today, a batch process requires multiple pump passes through the diafiltration membrane unit and uses a large amount of buffer. Zydney has been developing a continuous, countercurrent staged system that uses in-line dilution instead of a mixer. Diafiltration buffer is added to the product stream and membrane modules are used to remove buffer. Countercurrent staging, he said, significantly enhances buffer exchange by effectively re-using the buffer multiple times. Tests have shown that impurity removal remains constant throughout the process in a true steady state operation, and that the entire product "sees" the exact same process conditions, which offers the potential for improved product quality. He noted that he could design systems to produce a desired level of impurity removal simply by changing the number of stages and the flow rate of the diafiltration buffer relative to that of the feed buffer. A three-stage system, for example, can easily achieve a 1,000-fold removal of impurities and 99.9 percent buffer exchange, similar to what would be achieved in a batch system that uses more buffer.

Countercurrent staging can also be used in the chromatography step. In this case, it replaces a series of columns producing a cyclical rather than steady state response with a system that flows the chromatography resin as a slurry through a series of static mixers and hollow fiber membrane modules (see Figure 3). In this approach, which Zydney calls continuous countercurrent tangential chromatography, binding, washing, elution, and stripping are performed directly on the slurry, with the hollow fiber membranes and static mixers controlling separation and residence time. Countercurrent staging, he added, makes more efficient use of the buffer than would ever be achieved by mixing the slurry with the elution buffer, and it achieves purity levels that are "reasonably competitive" with multi-stage batch column configuration.

Continuous staged precipitation offers the opportunity to provide a low-cost method of product purification by using targeted precipitating agents such as metal chelators, volume exclusion agents, solvents, and affinity ligands. Continuous staged precipitation is the workhorse of the plasma fractionation industry, but it was never considered viable for most biotherapeutics when titers were in the one-tenth of 1 gram per liter range. Now that titers are pushing 10 to 20 grams per liter, it can be practical as an early capture step, replacing chromatography. Impurities would be removed using a countercurrent staged operation requiring small amounts of wash buffer.

Turning Concept into Reality

Mark Brower from Merck noted that the pipeline of biologics is evolving, which means that a new production platform must be flexible to be able to handle a variety of different types of products. He and his colleagues have been looking at various design concepts around single-use facilities that would afford the ability to respond to different product demand profiles and rely on various supply partners for the single-use components. The overriding philosophy would be to build out the facility as needed and at the right time.

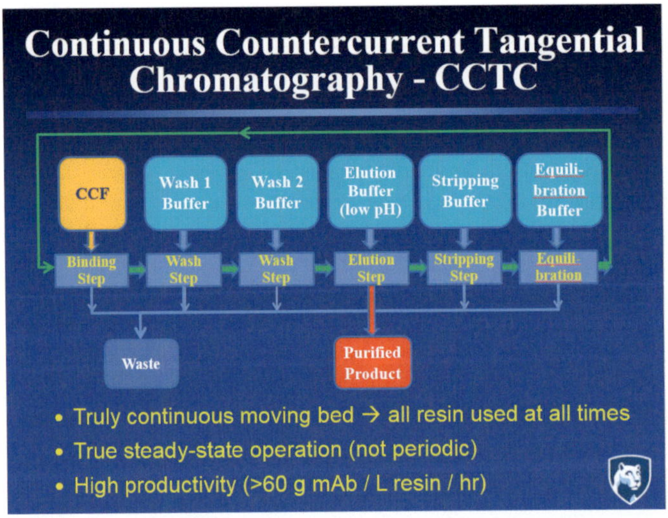

FIGURE 3 Conceptual diagram of a continuous countercurrent tangential chromatography product. NOTE: CCF refers to cell culture fluid, mAb refers to monoclonal antibodies. SOURCE: Zydney, slide 12.

To date, Brower's team has demonstrated that it can scale a continuous chromatography system and get consistent production of high-quality material, and it is now working to run the system over an extended period. The company has also built a pilot lab to test new system designs and learn how the different unit operations function. The lab includes every step of production from the bioreactor through final purification, all connected in a single-use, automated, closed system that is agnostic to the equipment within the supply chain. Two people, not including those doing media and buffer preparation, can run and monitor the entire system and change out consumables. The system includes process analytical tools that enable operators to intervene if the system deviates from the desired parameters. These tools also generate data that power multivariate modeling for process control and improvement.

The challenges Brower wants to address include validating the system, particularly during startup and shutdown. Run-time evaluation will look at real-time data to identify deviations and responses to those deviations. He is also concerned about how the organization deals with this new technology and how it responds to changes in the current development paradigm. There is also a number of technical challenges, including optimizing facility design for different operations, residence time distribution modeling, developing a control strategy for microbial agents, viral safety validation, demonstrating robust, process-scale good manufacturing process (GMP) systems, developing single-use sensors and consumables, and demonstrating connectivity using plug and play automation solutions.

PRODUCT MANUFACTURING

Drying Technologies to Stabilize Labile Molecules

Satoshi Ohtake from Pfizer reiterated Woodcock's message that FDA is encouraging industry to adopt novel technologies. As other speakers noted, there are both benefits and risks to adopting continuous manufacturing for biologics, and one question Pfizer has been asking is whether it should be looking at integrated, continuous manufacturing from start to finish or if it should consider piecemeal continuous manufacturing, which are two different value propositions. Toward that end, the company has been asking what Ohtake called a very simple set of questions:

- Is it feasible or necessary to implement continuous manufacturing?
- What are the available technologies and what is their readiness for implementation?
- When should these new technologies be implemented—at the time of new production introduction, as a part of life-cycle management, or as part of general technology development?

- How can it overcome the inevitable hurdles that come with new technology implementation?
- Who will take ownership of development and implementation?

Turning to the subject of drying process technologies, Ohtake said there is a great deal of knowledge to glean from other industries, including the food industry. The two approaches that have been applied to drug manufacturing are batch process freeze-drying and continuous process spray drying. Freeze-drying, or lyophilization, is the current gold standard and the bottom line is that it works and there are many applications for this technology. Freeze-drying, however, is expensive and time consuming, and there are issues with heterogeneity and scaling. With spray drying, product flow in equals product flow out, so scaling is not a problem.

One company in the food processing space is using microwaves instead of applied heat in the lyophilization process, and the company recently announced that it was collaborating with an equipment supplier to manufacture and deploy a continuous lyophilization system for use in pharmaceutical production. Another technology under development is spray freeze-drying, which generates frozen microspheres by dispersing substrate liquid using high-frequency nozzles into single droplets that then fall through a cooling zone and congeal into frozen spheres, which are then moved into a rotary vacuum dryer. This approach produces a narrow particle size distribution that is large enough to not be affected by static and can be filled into vials more easily. This instrument has been scaled from 1 to 2 liters per day to more than 100 liters per day. The one drawback is that the instrument's footprint is large. Ohtake noted that rotary vacuum drying is a batch process, so it would be necessary to couple droplet formation to multiple drying units to create a semi-continuous process. A third company is developing a continuous spray freeze-drying technology that replaces the rotary vacuum dryer with a vibrating, agitated drying chamber. Academic researchers, he said, are also developing continuous freeze-drying processes, one of which uses spin-freezing to create a thin layer on the outside of a drug vial filled with liquid product. The result is product coating the walls of the vial.

Ohtake said an important challenge for implementation is a lack of precedence, that is, nobody wants to go first when approaching regulators. The onus to work with regulators cannot be on the technology companies because they do not have the resources to do so. This needs to be a collaborative effort between industry and technology companies. Today, he said in summary, promising technologies are available, and FDA is supportive of innovative processing technologies. What is limiting progress is a lack of clarity on whether these technologies can be approved as platforms independent of a particular product that everyone can then use, or if the technology needs to be approved with a specific product. Also challenging is the prospect of introducing new technology as part of life-cycle management for an existing product as it would require changing the product license in every country in which it has been approved.

Solutions to Continuous Biomanufacturing Challenges

William Whitford from GE Healthcare said that there is an ongoing revolution in digital manufacturing brought on by the explosion in monitoring, analytics, artificial intelligence, automation, and advances in robotics. This revolution is enabling continual quality verification, product quality-based control of processes, and real-time release testing. Digital biomanufacturing consists of many disciplines (Whitford, 2017; Whitford and Julien, 2007), and one of the challenges for implementing it is that there are few people who are proficient in all of the component fields and the expertise often remains siloed, he said.

There are many information technology enablers of digital biomanufacturing, including adaptive systems employing artificial intelligence, the availability of big data and the effective management of large and complex data sets, and software suitable for FDA regulation. Whitford explained that digital biomanufacturing features include support for real-time prediction, analysis, and control of CQAs and critical process parameters. As a resident source of data, it also supports continuous optimization of process parameters and assists in the development and control of process intensification initiatives. Some of the performance goals for digital biomanufacturing include having a self-aware, continuously adaptive, autonomous plant monitored by remote experts. Such systems, said Whitford, are being implemented in other industries and are being considered for piecemeal adoption in biomanufacturing. Digital biomanufacturing should also support business continuity with incident control, management, and reporting capabilities, and in-line or on-line, real-time, orthogonal process monitoring and adaptive control. In short, digital biomanufacturing has the potential to support enterprise-level manufacturing intelligence in the form of reporting analytics for quality assurance and quality control support; monitoring analytics for process control, development, and optimization; and predictive analytics for scheduling, supply chain optimization, and optimal harvesting of product.

Digital biomanufacturing also fits well with next generation quality-by-design initiatives. One enabler of such efforts is multi-attribute methods of analysis, such as quadrupole Dalton mass spectroscopy, that can replace multiple traditional assays and identify process impact on multiple CQAs. Multi-attribute methods, said Whitford, are regulator friendly, cost-effective, and support both advanced process control and reporting on multiple product attributes in near real time.

Regarding monitoring solutions, Whitford said that automated at-line sampling technologies are commercially available, including on-line manifolds ready for in-line sensors, automatic at-line analytics, hands-off, contamination-free sampling. In fact, three companies now support cell-free sampling for multiplexed analysis from bioreactors and various downstream processes. Also available, he said, are various monitoring technologies for making specific types of continuous, real-time measurements. Single-use, adaptable in situ Raman probes, for example, can simultaneously measure glutamate, lactate, glucose, glutamine, ammonium ion, osmolality, and viable and total cell density. One company has reported that its technology can provide glycan analysis within 30 minutes (see Figure 4), while another uses nanofluidic chips to assay thousands of single cells in parallel and provide a point analysis of modal cellular characteristics rather than an average of all cells in a reactor. Whitford said he is excited by the possibility of using field-effect biosensing technology based on graphene to monitor discrete local changes in reactor conditions in real time.

At a larger scale, predictive control products that support multiplexed analysis, often in real or near real time, are commercially available. This type of technology is part of what GE Healthcare is working toward with its bioreactor digital twin project, a digital, mixed model, in silico representation of the bioreactor. Each implementation of the bioreactor digital twin would be specific for each bioreactor, updated with data from sister reactors as well as from the individual bioreactor. He noted that the continued advances in computing power open a wealth of possibilities for the future of digital biomanufacturing that cannot even be imagined.

Advanced process monitoring, using new sources of data and new analytical techniques, will lead to changes in process development and control. For example, being able to measure glycoform production in near-real time offers new approaches to using changes in product attributes for process control functions rather than using a representative value, such as pH or glucose levels. In addition, being able to monitor many divergent process parameters in both upstream and downstream processes offers the possibility of using this mass of data to control processes more efficiently.

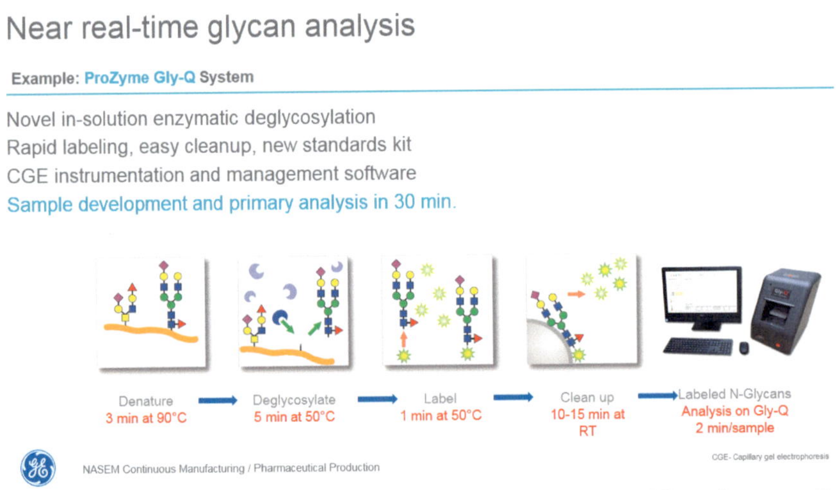

FIGURE 4 Diagram of the near real-time glycan analysis. NOTE: CGE refers to capillary gel electrophoresis, Gly-Q refers to complete integrated system for n-glycan analysis, RT refers to real-time. SOURCE: Whitford, slide 20.

Opportunities for Low-Cost Vaccine Manufacturing

Historically, said Tarit Mukhopadhyay from University College London, continuous manufacturing for vaccine manufacturing was rarely considered based on the small doses needed for vaccine efficacy, and when it was used, yields were low, requiring long production times to produce the needed material. However, a problem the vaccine industry faces is that vaccine manufacturing processes that were developed more than 50 years ago are largely unprofitable, which has led to many companies dropping out of the vaccine business. He noted that the Global Vaccine Action Plan, which the World Health Assembly endorsed in 2012 to increase global coverage for vital immunizations by 2020, is failing largely because of the unprofitability of vaccine manufacturing. In addition, because those vaccine manufacturers are operating at full capacity, any production glitch has an outsized effect on global availability of the affected vaccine.

The Bill & Melinda Gates Foundation is tackling this problem with its Innovations in Vaccine Manufacturing for Global Markets Grand Challenge, which seeks innovative approaches to producing low-cost vaccines with platforms suitable for 40 million doses annually at a target production cost of less than $0.15 per dose. The caveat to this challenge, said Mukhopadhyay, is that the core cost reduction cannot be realized primarily through economies of scale. His group, in collaboration with researchers from the Massachusetts Institute of Technology (MIT) and the University of Kansas, is tackling this challenge for a novel recombinant protein rotavirus vaccine.

To break the economies of scale model for vaccine manufacturing, it is necessary to realize cost savings through strain engineering and molecular design to improve product titer and quality, cut material and labor costs through simplified and intensified processes, and reduce facility-related costs and the overall footprint of the production process. Mukhopadhyay and his collaborators learned the first step was to engineer the antigen and select more productive cell lines, leading to a 4-fold increase in strain productivity (see Figure 5). They have also redesigned the antigen to remove glycosylation sites, producing a higher quality product. This approach of integrated molecular design improves product quality and productivity.

Turning to the manufacturing process itself, the team took advantage of the fact that engineered cells secrete the desired antigen, eliminating the need to harvest the cells, lyse them, and collect the product. This enabled the team to use a 50-liter perfusion reactor and couple it directly to the downstream processes, thereby reducing the labor component and the need for repetitive quality control testing, creating a highly intensified process through fill and finish. Overall, process yields are high enough to enable the entire process to fit inside a refrigerator-sized unit complete with automated process control. What this means, said Mukhopadhyay, is if three refrigerator-sized units can meet production needs, the process footprint can be 50-fold smaller and capital expenditures can be 15-fold less, reducing the barrier for entry for new companies.

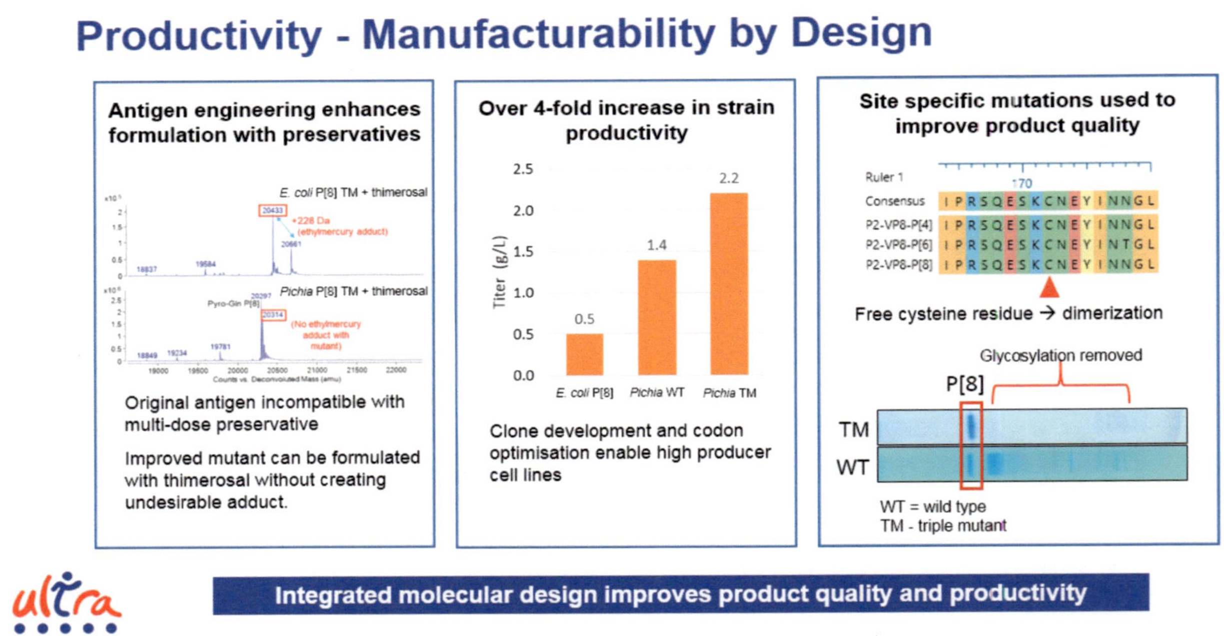

FIGURE 5 Diagrams displaying how engineering of the antigen can improve product quality. SOURCE: Mukhopadhyay, slide 6.

Modeling shows that a fully integrated process can not only meet the target for cost of goods, but also actually come in at an estimated 22 percent below that target. In a multi-product facility, cost of goods would be an estimated 36 percent less than the targeted $0.09 per dose. The benefits of moving to such an integrated process, however, could extend beyond profitability. Currently, vaccines are manufactured in centralized facilities requiring long distribution routes that lead to stagnant coverage and missed immunization opportunities. A distributive manufacturing network, based on an integrated and intensified process, would likely improve local and regional supply that could be tailored to meet demand and democratize vaccine supply. A distributed network would also increase manufacturing capacity overall and greatly increase the availability of vaccines to respond to emerging epidemics.

In closing, Mukhopadhyay said that applying the principles of continuous manufacturing to vaccines opens up new possibilities for increasing capacity and affordability. Realizing this promise, however, will require regulatory engagement to approve reduced quality control testing and real-time release based on the use of multi-attribute testing and improved in-process analytics. It will also require advocacy and regulatory support for a licensed product.

INTEGRATION

A Case Study: Biologically Derived Medicines on Demand

Christopher Love from MIT noted that integration is more than just connecting parts together, but rather involves syncing technology with the biology to be implemented in a continuous process. To him, mapping out the integration process before attempting to do it is imperative. The project his group has been working on is part of DARPA's Biologically Derived Medicines on Demand project, which envisions a different supply chain in which raw materials are transported around the world and the final product is made in a multi-product facility near the site of intended use. This approach would be useful for producing drugs to treat rare and orphan diseases, responding rapidly to outbreaks, and expanding global access to medicines.

Love defined integration as the act of combining or adding parts to make a unified whole in a way that blends or fuses the parts together so that the whole is greater than the sum of the parts, and his approach has been to embrace the entire design cycle at once. While he did not go into the details of the technological advances that his team made as part of this project, he noted that their development was driven by the notion of integrating the biology with the hardware (Matthews et al., 2017; Timmick et al., 2018). The result is a process for developing new molecules that in the best case takes approximately 12 weeks. This platform uses yeast as the production vehicle with a new type of media, developed using RNA sequencing and metabolomics, that is better defined, less expensive, and performs better than existing media. One advantage of working with yeast is that there are only 150 proteins in its secretome, an order of magnitude less than for CHO cells. Given that tractable number, it was possible to build a database for how a number of different resins interacted with each of these proteins, as well as drug products, under different conditions. This database enabled Love and his colleagues to create in silico algorithms for designing continuous, multi-column processes designed to work together in a flow-through sequence for any new molecule. Love's group also built a hardware system that enabled testing at scale and would be suitable for manufacturing, reducing technology transfer issues.

To date, his team has built three benchtop systems all operating with similar specifications. Each system comprises unit procedures: upstream, downstream, and formulation. One system produced clinical-grade G-CSF comparable to the FDA-approved product. He noted that his team used extensive gene sequencing, which is affordable because of the small size of the yeast genome, to characterize what is happening in the yeast cells during product production and use that to fix the yeast strain to be fit for purpose.

In short, said Love, deep biological knowledge has accelerated process development. This approach has enabled him to think about plug-and-play process development for new molecules, including human growth hormone, interferon-α2b, and rotavirus-specific nanobodies. For this last project, it took him and his team approximately four weeks from having the DNA sequence of the nanobody to producing six grams of product suitable for preclinical studies. Typically, the process takes 12 to 16 weeks from having a DNA sequence for the product to having material suitable for toxicology studies.

All of this information is building on itself, said Love, enabling his team to shorten the time for development and understanding what the production organism needs from a design perspective to maximize production. Integrating biological knowledge, he said, facilitates quality by design and has enabled his group to build a refrigerator-sized unit capable of producing up to 1 kilogram of product per year, which is equivalent to between 10 and 100 million doses of vaccine, 1 to 10 million doses of a cytokine, or 100 to 1,000 doses of a monoclonal antibody. His team has tested a single-use prototype, which currently sits at technology readiness level 5. In closing, Love said that integrated thinking is essential for achieving holistic designs and solutions, dematerializing and intensifying processes, and providing the agility to meet rapidly changing demand.

Strategy for Implementing Real-Time Release Testing

Richard Braatz from MIT recalled that an MIT-Novartis project several years ago had designed a plant-wide control system from first principles, built the fully integrated, end-to-end system, and showed that it reduced production costs by approximately 50 percent and met all purity specifications (Lakerveld et al., 2015; Mascia et al., 2013). That kind of system is continuing to move forward, he said.

Achieving the goal of having fully automated process development requires greatly increased understanding and optimization of each unit operation and exploiting process intensification, said Braatz. Fully automated high-throughput microscale technology can enable fast, continuous process research and development by generating a large amount of data to inform development of larger-scale systems. In addition, plug-and-play modules with integrated control and monitoring and corresponding in silico models will facilitate system deployment, as will dynamic models for unit operations for plant-wide simulation and control design. Smart data analytics and model-based control systems will help optimize operations, including startup, changeover, and shutdown. Addressing the design of control systems based on a virtual production facility, Braatz said it would ideally be built from first principles when possible using the highest complexity models available and be suitable for purpose for inventing and optimizing process designs and lower complexity models for process control design and quality monitoring (Lu et al., 2015).

Real-time release testing, said Braatz, requires being able to evaluate and ensure the quality of in-process and/or final drug product based on process data, which typically include a valid combination of measured material attributes and process controls (see Figure 6). Real-time release testing has the potential, he said, to increase quality assurance, increase yield by lowering rejection rates, and reduce cycle times. The key to enabling real-time release testing is the availability of integrated sensor technologies, mathematical models, and control strategies.

Braatz has utilized four strategies for ensuring that a particular CQA specification is satisfied (Jiang et al., 2017; Lee et al., 2015; Myerson et al., 2015), starting with direct measurement of CQAs. If that is not possible, the second strategy is to predict CQAs based on a first principles model powered by measurements of related variables and running in parallel with operations. The third strategy predicts CQAs based on a semi-empirical model, such as a response surface map or a partial least-squares model, powered by measurements of other variables, and the fourth strategy is to operate the critical process parameters so that they fall within a design space or a specified set of parameters shown in offline studies to provide assurance. The first three are applicable for close-loop feedback control strategies and the fourth is used for an open-loop strategy.

He noted that strategies to assure on-specification product have already been demonstrated for an end-to-end plant as far back as 2012 (Lakerveld et al., 2015). Braatz and his colleagues built first principles dynamic models for each unit operation, validated the models, and then placed them into a simulation to design unit operation and plant-wide controls. The first time the controllers were turned on for the real system, the resulting small molecule was on specification, he said.

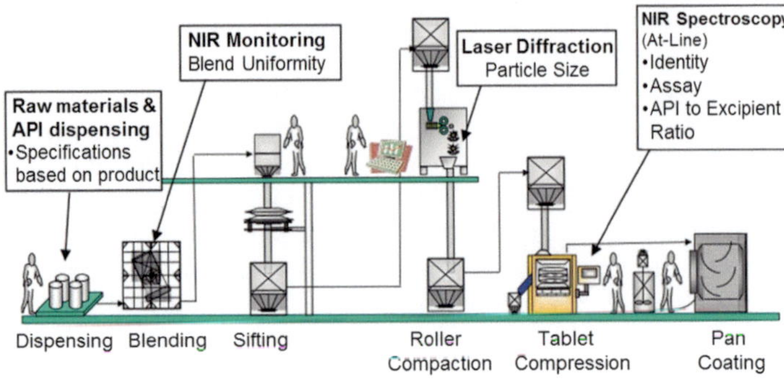

FIGURE 6 Conceptual diagram of real-time release testing. NOTE: API refers to active pharmaceutical ingredient, NIR refers to near infrared. SOURCE: Braatz, slide 11.

Advancements Toward Real-Time Release

Doug Richardson from Merck spoke about how he and his colleagues are addressing adaptive process control and process analytical technology to learn more about these processes in the near term and with the eventual goal of realizing real-time product release. Merck, he said, has a long history of developing and using process analytical technologies, which include both the analyzers and the system in which they fit. This process comprises four principles:

1. *Measure* process parameters and CQAs in a timely manner.
2. *Model* the process using information about the process conditions and product attributes to increase process understanding.
3. Using the improved process understanding to apply an appropriate *control* strategy to maintain the process in a state of control and product attributes within specifications.
4. Use improved process understanding to *optimize* process operation to ensure consistent attainment of CQAs and realize process efficiency improvements.

Richardson added that the earlier process analytical technologies are added to a process under development, the more information they will generate to improve process development and ease the transition into manufacturing.

Richardson explained the differences between at-line, on-line, and in-line process analytical technologies. At-line refers to the case where samples are collected manually and the analyzer is located next to the process. On-line systems are connected directly to the process and collect and automatically analyze samples, which are never returned to the process. In-line process analytical technology systems, the desired setup, are incorporated into the flow of the process and produce continuous data without sampling using capacitance, light scattering, spectroscopy, on-line liquid chromatography, and other types of sensors. Technology requirements for in-line systems include fast acquisition frequency, no required recalibration, an operation time of 2 weeks to 60 days, sterilizability, and the ability to monitor CQAs. He noted that his group has tested Raman devices, light-scattering technologies for real-time molecular mass measurements (Patel et al., 2018), and on-line ion exchange technologies to assess charge variants (Patel et al., 2017).

Today, it takes approximately 30 days to run the dozens of assays required for quality control leading to product release. To disrupt the quality-control process, Richardson and his team have been investigating peptide mapping mass spectrometry as a multi-attribute analytical tool (Rogers et al., 2015). This approach uses a streamlined mass spectrometry appropriate for quality control that has the potential for automated sample preparation and automated data processes. Currently, his group is using this tool to develop cell lines and to monitor in-process quality attributes. He noted that multi-attribute mass spectrometry performs comparably to individual assays and represents an orthogonal method for in-process control, release, and stability assessment. It is not a high-throughput approach, but it is high-value, sensitive, and selective. Richardson added that this multi-attribute method might also be able to provide a link to clinical data to help identify early in development those CQAs that are important for drug performance and perhaps streamline what needs to be tested for product release.

In his opinion, there is no path to real-time release without process analytical technologies, and mass spectrometry will likely play a role in getting to real-time release. He explained that the adoption path will be to measure one attribute at a time and add additional attributes with experience, and that even without the ability to measure everything, multi-attribute mass spectrometry has the potential to reduce the number of required assays for quality control. There are several technical challenges to address before process analytical technologies can enable real-time release, including the development of robust, single-use sensors, advanced data analytics and process modeling, and robust aseptic sampling and clarification.

In a panel discussion with all the session speakers, Charles Cooney, the discussion moderator, summarized some themes from the session: (1) integration is the start of the design, (2) innovate by choice not chance, (3) automation is not the end point; build from the start, (4) real-time release testing is an overarching strategy for integration, (5) modularity is important (of sensors and unit ops); plug and play, and (6) measure the minimum number of attributes; only test the essence of what is needed, not all of the attributes that it is possible to test.

REGULATORY AND QUALITY ASPECTS OF CONTINUOUS MANUFACTURING

Key Aspects of Regulation

Moheb Nasr from Nasr Pharma Regulatory Consulting recapped discussions held during two international symposia on continuous manufacturing that occurred in 2014 and 2016 at MIT. The outcome of the first symposium, which focused primarily on small molecule production, was a white paper on the regulatory and quality considerations for continuous manufacturing that established a regulatory baseline for continuous manufacturing (Allison et al., 2015). The second symposium, which split the discussion equally between small molecules and biologics, supported the need for a harmonized International Council for Harmonization of Technical Requirements for Pharmaceuticals for Human Use (ICH) guideline on technical and regulatory aspects. A second white paper resulted (Nasr et al., 2017). A third symposium took place in London on October 3-4, 2018, with a focus on case studies, business cases, and supply chain impact in both the small molecule and biologics spaces.

The currently regulatory environment, said Nasr, supports advancing innovation and the scientific and technical foundation on which regulations are based. Nasr noted that regulatory authorities within ICH, as well as an increasing number of non-ICH regulators, are encouraging industry to adopt new technologies. Recent ICH guidelines, in fact, emphasize science- and risk-based approaches for quality assurance. He added that regulatory expectation of assurance of reliable and predictable quality is much the same for both batch and continuous manufacturing and that a proposed ICH guideline, called ICH Q13, would establish appropriate standards to facilitate harmonized global implementation for continuous manufacturing.

The current regulatory framework, said Nasr, is generally adequate to support continuous manufacturing, but traditional concepts do need to be explored further or challenged to advance continuous manufacturing. Early and frequent communication between manufacturers and regulators is encouraged, he noted, and FDA's Emerging Technology Team has played a key role here. In-process controls and sampling considerations are different than batch processes and should be established accordingly, and he added that the field needs to develop and define acceptable procedures for handling process deviations. Nasr noted, too, that a control strategy would be expected to assure regulators that the manufacturing process consistently produces product of the desired and intended quality.

In continuous manufacturing, special considerations include the state of control, raw materials and intermediates, equipment, product collection or rejection, traceability, process monitoring and sampling, and specifications. Regarding the state of control, Nasr explained that maintaining a state of control provides assurance of consistent and desired product quality, which means that the control strategy should have the ability to detect process upsets and institute corrective actions to bring the process back into conformance. For raw materials and intermediates, it may be necessary to have additional controls when multiple lots of a raw material are used in a single continuous manufacturing batch. Special maintenance, calibration, and periodic review of continuous manufacturing equipment are also needed to ensure the batch remains within operating specifications over the duration of the continuous manufacturing process.

Other special considerations include the need to establish a priori criteria, based on level of risk, for product collection and product rejections, including rejection of an entire run. It is also necessary, said Nasr, to document traceability of income materials to the final product. He noted that the purpose of the monitoring system is to manage planned changes and respond to unplanned disturbances in the process.

There are also quality and GMP considerations, said Nasr. He advised using current GMP processes to guide the implementation of a continuous manufacturing pharmaceutical quality system, with assessments and revisions to the current process as needed rather than starting from scratch. Such an approach, he said, will make regulatory inspections go more smoothly and require less time to complete, as opposed to developing an entirely new standard operating procedure for quality control and assurance. Other considerations include detailing startup and shutdown procedures, how production collection and in-process sampling will occur as a means of assuring continued process performance and product quality, process validation and continued process verification procedures, material traceability, personnel and training procedures, and how cleaning will be validated.

To bridge an existing batch manufacturing process to a continuous process, the continuous process can be introduced as a new process for a new molecular entity or as a post-approval manufacturing change. For the latter approach, it will be necessary to establish that the product is physiochemically equivalent as it is produced by the continuous process. For low-risk changes to product CQAs, such as polymorphicity, dissolution, impurities, and stability, demonstration of chemical equivalence could be sufficient to support the change from batch to continuous. For high-risk changes, such as significant formulation changes or drug release characteristics, bioequivalence studies may be needed.

In conclusion, said Nasr, regulators are supportive of pharmaceutical manufacturing innovation, but this type of forum should be used to be explicit about current regulations that may impede the adoption of continuous manufacturing. Mentioning the 2014 white paper, he restated that it provides a good foundation for implementing continuous manufacturing now, though there is a need for risk-based assessments and quality-by-design tools, including process analytical technologies, to drive the development and implementation of continuous manufacturing. Nasr also noted the need to evaluate pharmaceutical quality systems and relevant standard operating procedures to assure successful implementation and compliance with current GMP. He ended his presentation by calling on the workshop participants to review and comment on the proposed ICH Q13 guideline, which he believes provides a great opportunity to improve regulatory standards.

Considerations for Virus Clearance Validation

Lisa Connell-Crowley from Just Biotherapeutics said that viral safety for biotherapeutics produced in mammalian cells is enshrined in ICH guideline Q5A and calls for virus testing of the cell bank, virus testing of the process bulk production, and demonstration that the downstream process can clear viruses. In terms of the downstream process, it is important to demonstrate robust and effective viral clearance of potential undetected adventitious virus and noninfectious retrovirus-like particles (RVLPs) expressed by CHO cells or other rodent cell lines. Viral clearance is provided by two dedicated, orthogonal steps—usually low-pH viral inactivation and viral filtration, explained Connell-Crowley. In some processes, additional clearance occurs during chromatography steps between inactivation and filtration. Viral clearance is assessed via spiking studies with between two and four model viruses using scaled-down models run at target or worst-case scenarios. Most companies have rule-of-thumb clearance targets they use.

In batch production, viral clearance assessment is a straightforward process where the load material of a batch unit operation is spiked with a model virus and then virus levels are measured in the resulting pool to determine the clearance of virus by that unit operation. This approach does not work for continuous manufacturing, as it will not determine how well an individual step removes a virus and it is likely to underestimate viral clearance, said Connell-Crowley. What she and her colleagues do with a homogeneous load is to quantify viral clearance for each discrete step in the continuous process. For non-homogeneous load, her team implemented in-line virus spiking using a push-pull syringe to spike a virus into the feed stream of a second unit of a two-unit operation. This approach, she said, works well for the chromatography and filtration steps.

Regarding continuous capture chromatography, evaluating virus clearance is difficult for a situation in which multiple columns are overloaded and product breakthrough on column one is directed to the second column, and so on. She suggested evaluating virus clearance using a virus-spiked load and flow-through material on a single bench column scale or to test clearance of RVLP in the load using a quantitative polymerase chain reaction (qPCR) instead of performing spike studies using a model virus.

Low-pH virus inactivation of enveloped viruses can be done in a batch mode, using discrete acid titration in a mixed holding tank or with a continuous system using a coiled flow reactor and in-line acid titration (Orozco et al., 2017). Continuous viral filtration has its own challenges, particularly the extended time for filtration in a continuous process versus a batch process. Continuous filtration processes operate at a constant low flow to optimize process time and filter utilization while minimizing operator manipulations and cost. The result load, said Connell-Crowley, can be homogeneous or variable depending on the upstream process.

Virus Spiking is a Challenge for a Long Duration, High Load Continuous Viral Filter Step

- Bracketed spiking: Test the integrity of the filter periodically by introducing virus spike at defined times.
  - Beginning and end? Once per day?
- Variable bracketed: Same concept as bracketed but includes a low level virus spike throughout duration when not using high virus spike
- Other options?

Spiking Strategy	Virus Spike %	Virus load (log_{10} $TCID_{50}$)	Pool LRV*
High virus spike	1% in 720mL	8.48	≥ 5.94
Low virus spike	0.01% in 720mL	6.17	≥ 3.64
Bracketed	0.5% in 30mL, 0% in 60mL, 0.5% in 30mL	7.20	≥ 5.80*
Variable bracketed	0.5% in 30mL, 0.003% in 60mL, 0.5% in 30mL	6.79	≥ 5.01*

*Average of the 0.5% bracketed spikes, which were within 1 log of each other

Just.

FIGURE 7 Challenges with virus removal in different spiking scenarios. SOURCE: Connell-Crowley, slide 14.

Regarding assessment, virus spiking is a challenge for a long duration, high-load, continuous virus filtration step, she explained (see Figure 7). Loading too much of a virus on a filter can lead to non-representative filter fouling from virus preparation contaminants and a virus breakthrough that is not representative of a contamination event. Loading too little of a virus on a filter can result in numbers that are lower than the target clearance value because of lower assay sensitivity. One alternative strategy is to use bracketed spiking that periodically tests the integrity of the filter by introducing a virus spike at defined times. Another strategy is to use variable bracketed spiking that also includes a low-level virus spike throughout the duration of the filtration process when not using the high virus spike.

For RVLP quantification for continuous manufacturing processes, Connell-Crowley said the challenge is to determine where to sample for RVLPs to determine the RVLP burden. If, for continuous perfusion processes, the RVLPs are retained by the perfusion filter while the product passes through, the question becomes whether the best place to look for them is in the bioreactor or the permeate. She noted that her team is using qPCR to measure RVLPs because it is more sensitive, faster, and more cost effective than the standard technique of using transmission electron microscopy. Using qPCR, she and her colleagues found that there are high levels of RVLPs in the bioreactor and three to four logs lower levels in the permeate, which feeds into downstream processes, with three different perfusion membranes.

In conclusion, Connell-Crowley said that the challenges of continuous downstream processes require rethinking how viral clearance assessments are done given that current virus spiking strategies will not necessarily work in all situations. The goal, she said, is to ensure viral safety without constricting process design because of concerns about difficulty of viral validation.

Considerations for Integrated Biomanufacturing

In the workshop's last presentation, Veena Warikoo from Roche noted that there will not be a good business case for continuous manufacturing without process intensification. Process intensification as chemical engineers have defined it is the development of novel apparatuses and techniques that are expected to bring dramatic improvements in manufacturing and process, and by doing so, substantially decrease the ratio of equipment size to production capacity, energy consumption, or waste production with the result of creating a cheaper, more sustainable process. Process intensification does not just replace an old, inefficient plant with new, intensified equipment. Rather, it can challenge business models, opening opportunities for new patentable products, process chemistry, and change to just-in-time or distributed manufacturing. Continuous manufacturing, said Warikoo, is just one tool for achieving intensification, and business drivers dictate the degree of intensification.

Among the most frequently discussed challenges to an intensified process are that it is not needed, that there would be regulatory challenges, it was too complex, the virus clearance would be impossible, and handling perturbations would be too difficult. As the presentations at this workshop have shown, those challenges have proven to be solvable. For example, since the 1990s, when the field of continuous pharmaceutical manufacturing was first implemented, to today, manufacturing of monoclonal antibodies has evolved from state-of-the-art large fed-batch and batch processes to intensified, semi-continuous processes. Given where the community is, she predicted that it will eventually develop a fully continuous process for producing monoclonal antibodies and other biologics and reap the associated benefits that other industries have realized from continuous manufacturing.

Warikoo reiterated Woodcock's message that FDA is taking numerous steps to promote continuous manufacturing to improve quality in the pharmaceutical production process. In 2015, FDA issued guidance on the matter titled *Advancement of Emerging Technology to Modernize the Pharmaceutical Manufacturing Base*. In addition, FDA's most recent budget request included an additional $400 million to support planned initiatives aimed at supporting new and ongoing efforts to foster more investment and innovation in the development of therapeutics and diagnostics, including the movement toward continuous manufacturing as a means to improve the agility, flexibility, cost, and robustness of manufacturing processes. She also said that ICH Q13 will include key definitions, scientific principles, control strategies, validation strategies, and regulatory expectations as they pertain to continuous manufacturing.

She believes that risk-based compliance will change little for continuous manufacturing, in large part because the basic technologies for both upstream and downstream processing have not changed other than to make them connectable to each other. The one place where new understanding will be needed in terms of regulatory issues concerns the effects of flow on product quality. She also said that dynamic operability as a control strategy is a must-have feature going forward, in her opinion.

The quality risk management process, she added, will not change when going from batch to continuous manufacturing, but what will change is the control strategy. The unique challenges for developing a control strategy for continuous manufacturing include the fact that some processes, such as chromatography, will have a continuous feed but periodic output, and that while individual steps can be fast, the overall process will run for 30 to 60 days. There will also be fewer unit operations in an intensified process, but more columns and column cycles, and while the operation of the system is simple, the instrumentation required will be complex.

With regard to dynamic operability, both in-process product quality monitoring and process parameter monitoring are essential, said Warikoo. Statistical models can be built using the wealth of data produced by process analytical technology to establish correlations between process parameters, in-process product characteristics, and CQAs for the drug substance. With strong correlations, real-time release of the drug substance is possible, she added. Experiments she and her collaborators conducted showed that this approach is feasible, and that process analytical technologies can address process control challenges for integrated, continuous biomanufacturing. She explained that process analytical technologies provide real-time process monitoring, real-time process control, and automation with redundancies. Liquid flow rates through all unit operations are controlled continuously so that they are synchronous as a means of mitigating deviations.

REPORTS FROM SIX BREAKOUT SESSIONS

At two times during the workshop, participants broke into six small groups to answer questions posed by the workshop planning committee on specific aspects of continuous manufacturing. In the workshop's last session, a rapporteur summarized each small group's discussions.

Breakout Group 1: The Business Case

Question 1: What are some major economic challenges to moving forward with continuous manufacturing?

Gintaras V. Reklaitis of Purdue University described the discussion of the business case breakout session. According to Reklaitis, the group discussed that if all else is equal, continuous manufacturing will generally require lower capital expenditures and operating costs than batch manufacturing when the process operates as designed, though many processes are modified from the as designed mode when implemented. In several reported instances, capital expenditures and operating costs for continuous manufacturing have been 50 percent lower. Another possible advantage of continuous manufacturing is that it can make the supply chain more agile, though calculating the economic benefit of that is challenging.

A key challenge to adoption that was discussed is the real or perceived technical risk that the process will not be able to operate as designed, which gets at the robustness of the technology, the probability of failure of the line or loss of a product run, or more importantly in the case of a new product, a delay in that product's launch or loss of market position. This aspect of risk was seen to be a bigger barrier than the economics. Regulatory uncertainty is not perceived to be the primary risk given FDA's support for innovation, but this may be different outside of the United States.

Question 2: Discuss ideas that could help address the identified challenges. Are there new conceptual ideas that could improve the business case for continuous processes?

According to Reklaitis, many breakout session participants believed reduction of risk requires gaining in-house experience, but given the time it takes to bring new technology on line, that experience ought not to be gained on a new product. The strategies to adopt, then, that the breakout session participants discussed are to either share risk by participating actively in consortia, such as NIIMBL, or to convince corporate management to accept the risk for strategic reasons. Arguments posed by breakout session participants that could be used to convince management to accept the risk include that continuous manufacturing may provide a competitive advantage through gains in speed, reduction in cost, improvements in quality, and flexibility and therefore may be a tool for the organization's manufacturing operations. Other arguments posed for continuous manufacturing are that first adopters have already left the gate and that there may be competitive advantages when technology and know-how are in place when the right product comes along. Learning from other industries that have adopted continuous manufacturing can build confidence in this new process for manufacturing drugs. Another points discussed that may help management accept risk is framing a business case around a portfolio of products.

Question 3: Which product types will be the easiest and the hardest to make a business case for regarding continuous manufacturing?

The most favorable products would be those for which the manufacturing choices are limited to begin with, such as highly labile products, and those for which continuous components are currently needed, such as with perfusion-based products. Product families where deep knowledge exists in an organization would also fall into this category, because deep knowledge is needed for continuous manufacturing to succeed, as would biosimilars for which cost is a differentiator. Less favorable products would be cell and gene therapies, at least in the mid-term.

Breakout Group 2: Upstream Processing

Question 1: What are the major challenges to moving continuous manufacturing forward within upstream processing?

Ken Lee of MedImmune discussed some of the points made in the upstream processing breakout session. One point made by the group was that the equipment used for upstream processing was developed for batch processing, not perfusion processing, and differences in oxygen perfusion, mass transfer, and other aspects of existing equipment typically need to be modified for continuous upstream processing. Related to the perfusion issue is the fact that at high density or at higher growth rates, cells can quickly crash because of deficiencies in nutrient supply. While perturbations are to be expected, new scaling factors resulting from the increased demand of cells can lead to unanticipated fluctuations that the system cannot respond to in time to save the cells. In addition, nutrient demands may vary at different stages of perfusion culture.

Question 2: What are some research topic areas or ideas that could help address the identified challenges?

According to Lee, one of the ideas discussed by the breakout group is that designing a small number of media to meet different needs at different phases of growth would help address some of these challenges, as would the ability to collect data in different forms, such as in-line, at-line, on-line, and off-line, that could power modeling activities to determine which data and which times are valuable for identifying control levers. At a more fundamental level, understanding cell biology at high densities and how cell density affects cell nutrient consumption is important for designing control strategies and for determining why media are essential for performance.

Question 3: Are there mechanisms that could address a specific challenge, such as a change in regulation, a shift in organizational policies, or even a targeted short-term research contest, such as a challenge or a code-a-thon?

According to Lee, the breakout group did not discuss this question much. The main idea discussed was that companies need changes to their internal regulatory thinking. For example, by modifying how they perceive process changes fitting into the regulatory framework, companies may be less resistant to change and the regulatory burden associated with new filings.

Miscellaneous thoughts to come out of the discussion regarding upstream processing:

According to Lee, this group also discussed the development of new sensors for analytics and process properties that are currently unmeasured, such as osmolality. The group also discussed the use of frozen seed trains to get to the production vessel more quickly. It will also be important to determine the amount of product that attributes flexibility compared to batch attributes that regulators will allow to accommodate the benefits of continuous manufacturing. Increasing the level of automation in cell culture has the potential to reduce cell culture variability and increase consistency, and utilizing the literature on cell metabolism could provide new approaches to process control. Training will likely be important to enable the workforce to operate continuous manufacturing systems.

Breakout Group 3: Downstream Processing

Question 1: What are some major challenges to moving continuous manufacturing forward within downstream processing?

Eva Gefroh of Just Biotherapeutics reported back from the downstream processing breakout session. According to Gefroh, participants in this session discussed that implementing truly continuous downstream processes is more challenging than for upstream processing, in part because of the large number of unit operations in downstream processing and the challenges of integrating and controlling those operations. On the 5-year horizon, most of the technologies available or being developed are modifications of those used in batch mode, such as multi-column chromatography, while new technologies designed specifically for continuous downstream processing, such as crystallization and countercurrent technologies, will be developed over a 10-plus-year horizon. Separation challenges include fouling of membranes and resin surfaces, while cost challenges related to single-use operations include a shift from capital expenditures to the cost of consumables. Consumable costs may fall in the future as more vendors innovate in this area and as more people adopt continuous processes. By improving systems and models for data analysis, interpretation of the wealth of data generated by continuous processes will allow for better monitoring of system processes.

Question 2: What are research topic areas or ideas that could help address the identified challenges?

According to Gefroh, the group determined that three key areas of research focus for downstream processing include developing better process analytical technologies, predictive and self-learning models, and novel continuous processes; for example, finding a game-changing equivalent to Protein A for continuous processing, identifying new affinity ligands, developing alternate hosts that may allow novel processes to work, and designing new sorbents or membrane surfaces that are more selective for product rather than impurities.

Question 3: Are there other mechanisms that could address a specific challenge, such as a change in regulation, a shift in organizational policies, or even a targeted short-term research contest such as a challenge or a code-a-thon?

According to Gefroh, the group suggested that collaborating with national laboratories and participating in consortia such as NIIMBL will help with technologies that are close to being ready to implement, as well as new mechanisms for researching disruptive manufacturing technologies at technology readiness levels 1-3. Workshops such as this event will continue to be important to encourage further discussion and networking and to help dispel myths about continuous manufacturing, both from a technical and regulatory perspective. Private funding can also help further these activities.

Breakout Group 4: Product Manufacturing

Question 1: What are the major challenges to moving continuous manufacturing forward within product manufacturing?

Michael Ladisch of Purdue University moderated the product manufacturing session. According to Ladisch, some technological challenges discussed by the group include drying and mass limitations and the fact that newer product types, particularly vaccines, may not be amenable to freeze-drying and some vaccines may not be injectable after being subjected to novel drying techniques. To make vaccines affordable, the incentive will be to keep the cost of producing subunit vaccines low and stable.

Question 2: What are the research topic areas or ideas that could help address the identified challenges?

According to Ladisch, the group discussed the importance of getting industry involved so that industry can identify areas where research is needed and lead the research along with national laboratories and universities. The first thing to look at is the flexibility of manufacturing a vaccine and where the site of manufacturing will be. Smaller modular units might be appropriate for manufacturing subunit vaccines outside of the United States, for example. In addition, it will be necessary to develop a basic unit of operation for continuous manufacturing that is generally applicable to biologics, both vaccines and therapeutic proteins. Once in hand, those unit operations could be applied to situations such as a pandemic where a rapid ramp-up of production is needed.

Question 3: Are there other mechanisms that could address a specific challenge, such as a change in regulation, a shift in organizational policies, or even a targeted short-term research contest such as a challenge or a code-a-thon?

According to Ladisch, one mechanism discussed would be to explore how FDA, NIH, and BARDA could facilitate integration of microbial metabolomic and proteomic data across different companies to improve models. It was mentioned that it may be important, though, to preserve freedom to operate after sharing data.

Breakout Group 5: Integration

Question 1: What are the major challenges to moving continuous manufacturing forward within integration?

Charles Cooney of MIT moderated the breakout session on integration. According to Cooney, the group discussed open-source platforms, models, and plug-and-play unit operations that may encourage research and development. Such open-source platforms and models have the potential to accelerate regulatory approval. However, many in the group felt that achieving real-time release—including release of raw material and drug substance—is a non-trivial challenge requiring collaboration among many parties to develop technologies for implementing quality control and quality assurance, as well as a robust quality management system and quality management information system. The group also discussed that understanding deviations, both in the manufacturing process and CQAs, is an important challenge to overcome and this will be aided by mechanistic models that can drive a deeper understanding of the manufacturing process and how it affects the biology, chemistry, and physics of the product. One value of the growth of the biosimilar industry is that it is forcing the field to revisit and think deeply about what a clinically relevant CQA means. An additional challenge will be to show that a process consistently meets expectations, regardless of whether it operates in a steady state or not.

Question 2: What are research topic areas or ideas that could help address the identified challenges?

According to Cooney, the group discussed that from a manufacturing perspective, the idea of plug-and-play is attractive, and open source platforms will facilitate licensing. Cybersecurity may be a concern, and open access as a starting point can save time.

Breakout Group 6: Regulatory and Quality Aspects of Continuous Manufacturing

Question 1: What are the regulatory concerns regarding continuous manufacturing?

Keith Roper of the University of Arkansas moderated the breakout session of regulatory and quality aspects of continuous manufacturing. According to him, the group discussed that currently, ICH includes Australia, Brazil, Canada, China, the European Union, Japan, Korea, Singapore, Switzerland, and the United States. The symposium held in October 2018 in London[3] continued to accelerate the discussion on ICH Q13, which is expected to be completed and released in 2 to 3 years. In the United States, FDA's emerging technology team has stated that it welcomes conversations with industry and other potential groups about the technologies needed for continuous manufacturing, and that it holds many of these discussions each year. The European Medicines Agency has a process analytical technology team that has expanded its interests to include quality by design and continuous manufacturing. Japan has an innovative manufacturing technology working group and a forum for academia, industry, and regulators to hold ongoing discussions about continuous manufacturing. In addition, Japan has issued a white paper with provisional draft guidance on continuous manufacturing applied to small molecules. Other countries have shown an interest in hearing about new technologies.

---

[3]See https://www.iscmp2018.org (accessed December 17, 2018).

According to Roper, the group discussed that FDA's perspective on process modeling and multivariate assessment of product quality is that companies can design a process and provide the rationale and data that guarantee that the process is in a state of control to ensure product quality. Industry's concern rests with how much detail of process and engineering development will be subject to regulatory review and approval. For example, there were early instances when companies were apprehensive that there would be residual risk and expressed concern regarding design space submissions, so submissions were restricted to the operational space. At issue here is that this will hinder the development of continuous manufacturing from a regulatory perspective.

Roper explained that many in the group felt that regarding model quality and update issues, there is a need to have routine updating and maintenance of models, which is challenging in the current regulatory environment. It has been reported that the European Medicines Agency has concerns in this area and that FDA's draft guidance is rather strict regarding model updates. According to Roper, some felt that the role of modeling in development versus process control may be addressed in the upcoming ICH Q13, and requirements for updating models for continuous manufacturing may be lowered to notification only. Soft sensors, which are models that infer attributes that are not directly measurable, are updated already because they are considered in monitoring. Currently, companies determine how to ensure quality and how much information to include in a submission.

Other concerns Roper said the group discussed include the need for adequate and representative process analytical technology control sensor probes to ensure adequate sampling frequency. This is a technical issue for product design, not necessarily for a regulatory submission. A possible solution discussed for the challenges reported by Roper is that companies can engage regulators when they have questions, and newer ICH regions also want to know more about new technologies even if they do not have analogous teams. Roper said that many felt that engineering design and process development should not be subjected to too much oversight, but rather quality and control strategy should be the focus of regulators.

Roper also shared that many in the group felt that for ICH Q13, more input is needed on performance-based approaches and their suitability for continuous manufacturing, which may be why this was a key topic at the October 2018 symposium.[4] Roper also said that many in the group felt that ICH Q13 should have a role in laying the groundwork for global regulatory convergence because divergence on fundamental models and feedback control would make it more difficult for companies to comply with all of the international regulations.

Two additional concerns included a regulatory question on carryover from lot to lot. Single use technologies provide no carryover, but FDA has communicated that it expects companies to follow rules under 21 CFR 211 to determine

[4]See https://www.iscmp2018.org (accessed December 17, 2018).

which carryover is from a traceability standpoint. Under those rules, the company determines how to define the lot to see whether changes in product quality have occurred. The decision to collect, reject, or hold a product is then subject to further quality control information.

The second concern regarded the lifetime of components such as resins and filters. For batch processes, the components are used for several cycles and then virus spiking is done to validate that the membrane is clean and the process worked as designed. With continuous manufacturing, there is a question about how frequently validation would be required.

REFERENCES

Allison, G., Y. T. Cain, C. Cooney, T. Garcia, T. G. Bizjak, O. Holte, N. Jagota, B. Komas, E. Korakianiti, D. Kourti, R. Madurawe, E. Morefield, F. Montgomery, M. Nasr, W. Randolph, J.-L. Robert, D. Rudd, and D. Zezza. 2015. Regulatory and Quality Considerations for Continuous Manufacturing May 20-21, 2014 Continuous Manu-facturing Symposium. *Journal of Pharmaceutical Sciences* 104(3):803-812. doi: https://doi.org/10.1002/jps.24324.

Jacoby, R., L. Pernenkil, S. Harutunian, M. Heim, and A. Sabad. 2015. Advanced Biopharmaceutical Manufacturing: An Evolution Underway. Deloitte. New York, NY.

Jiang, M., K. A. Severson, J. C. Love, H. Madden, P. Swann, L. Zang, and R. D. Braatz. 2017. Opportunities and challenges of real-time release testing in biopharmaceutical manufacturing. *Biotechnology and Bioengineering* 114(11):2445-2456. doi: https://doi.org/10.1002/bit.26383.

Konstantinov, K. B., and C. L. Cooney. 2015. White Paper on Continuous Bioprocessing: May 20-21 2014 Continuous Manufacturing Symposium. *Journal of Pharmaceutical Sciences* 104(3):813-820. doi: https://doi.org/10.1002/jps.24268.

Lakerveld, R., B. Benyahia, P. L. Heider, H. Zhang, A. Wolfe, C. J. Testa, S. Ogden, D. R. Hersey, S. Mascia, J. M. B. Evans, R. D. Braatz, and P. I. Barton. 2015. The Application of an Automated Control Strategy for an Integrated Continuous Pharmaceutical Pilot Plant. *Organic Process Research & Development* 19(9): 1088-1100. doi: https://doi.org/10.1021/op500104d.

Lee, S. L., T. F. O'Connor, X. Yang, C. N. Cruz, S. Chatterjee, R. D. Madurawe, C. M. V. Moore, L. X. Yu, and J. Woodcock. 2015. Modernizing Pharmaceutical Manufacturing: from Batch to Continuous Production. *Journal of Pharmaceutical Innovation* 10(3):191-199. doi: https://doi.org/10.1007/s12247-015-9215-8.

Lu, A. E., J. A. Paulson, N. J. Mozdzierz, A. Stockdale, A. N. F. Versypt, K. R. Love, J. C. Love, and R. D. Braatz. 2015. Control systems technology in the advanced manufacturing of biologic drugs. IEEE Conference on Control Applications (CCA), September 21-23, 2015.

Mascia, S., P. L. Heider, H. Zhang, R. Lakerveld, B. Benyahia, P. I. Barton, R. D. Braatz, C. L. Cooney, J. M. B. Evans, T. F. Jamison, K. F. Jensen, A. S. Myerson, and B. L. Trout. 2013. End-to-End Continuous Manufacturing of Pharmaceuticals: Integrated Synthesis, Purification, and Final Dosage Formation. *Angewandte Chemie International Edition* 52(47):12359-12363. doi: https://doi.org/10.1002/anie.201305429.

Matthews, C. B., C. Wright, A. Kuo, N. Colant, M. Westoby, and J. C. Love. 2017. Reexamining opportunities for therapeutic protein production in eukaryotic microorganisms. *Biotechnology and Bioengineering* 114(11):2432-2444. doi: https://doi.org/10.1002/bit.26378.

Myerson, A. S., M. Krumme, M. Nasr, H. Thomas, and R. D. Braatz. 2015. Control Systems Engineering in Continuous Pharmaceutical Manufacturing May 20-21, 2014 Continuous Manufacturing Symposium. *Journal of Pharmaceutical Sciences* 104 (3):832-839. doi: https://doi.org/10.1002/jps.24311.

Nasr, M. M., M. Krumme, Y. Matsuda, B. L. Trout, C. Badman, S. Mascia, C. L. Cooney, K. D. Jensen, A. Florence, C. Johnston, K. Konstantinov, and S. L. Lee. 2017. Regulatory Perspectives on Continuous Pharmaceutical Manufacturing: Moving From Theory to Practice: September 26-27, 2016, International Symposium on the Continuous Manufacturing of Pharmaceuticals. *Journal of Pharmaceutical Sciences* 106(11):3199-3206. doi: https://doi.org/10.1016/j.xphs.2017.06.015.

Orozco, R., S. Godfrey, J. Coffman, L. Amarikwa, S. Parker, L. Hernandez, C. Wachuku, B. Mai, B. Song, S. Hoskatti, J. Asong, P. Shamlou, C. Bardliving, and M. Fiadeiro. 2017. Design, construction, and optimization of a novel, modular, and scalable incubation chamber for continuous viral inactivation. *Biotechnology Progress* 33(4):954-965. doi: https://doi.org/10.1002/btpr.2442.

Patel, B. A., N. D. S. Pinto, A. Gospodarek, B. Kilgore, K. Goswami, W. N. Napoli, J. Desai, J. H. Heo, D. Panzera, D. Pollard, D. Richardson, M. Brower, and D. D. Richardson. 2017. On-Line Ion Exchange Liquid Chromatography as a Process Analytical Technology for Monoclonal Antibody Characterization in Continuous Bioprocessing. *Analytical Chemistry* 89(21):11357-11365. doi: https://doi.org/10.1021/acs.analchem.7b02228.

Patel, B. A., A. Gospodarek, M. Larkin, S. A. Kenrick, M. A. Haverick, N. Tugcu, M. A. Brower, and D. D. Richardson. 2018. Multi-angle light scattering as a process analytical technology measuring real-time molecular weight for downstream process control. *mAbs* 10(7):945-950. doi: https://doi.org/10.1080/19420862.2018.1505178.

Pollock, J., J. Coffman, S. V. Ho, and S. S. Farid. 2017. Integrated continuous bioprocessing: Economic, operational, and environmental feasibility for clinical and commercial antibody manufacture. *Biotechnology Progress* 33(4):854-866. doi: https://doi.org/10.1002/btpr.2492.

Rogers, R. S., N. S. Nightlinger, B. Livingston, P. Campbell, R. Bailey, and A. Balland. 2015. Development of a quantitative mass spectrometry multi-attribute method for characterization, quality control testing and disposition of biologics. *mAbs* 7(5):881-890. doi: https://doi.org/10.1080/19420862.2015.1069454.

Timmick, S. M., N. Vecchiarello, C. Goodwine, L. E. Crowell, K. R. Love, J. C. Love, and S. M. Cramer. 2018. An impurity characterization based approach for the rapid development of integrated downstream purification processes. *Biotechnology and Bioengineering* 115(8):2048-2060. doi: https://doi/org/10.1002/bit.26718.

Walther, J., R. Godawat, C. Hwang, Y. Abe, A. Sinclair, and K. Konstantinov. 2015. The business impact of an integrated continuous biomanufacturing platform for recombinant protein production. *Journal of Biotechnology* 213:3-12. doi: https://doi.org/10.1016/j.jbiotec.2015.05.010.

Whitford, W. 2017. The Era of Digital Biomanufacturing. *BioProcess International* 15(3):12-18.

Whitford, W., and C. Julien. 2007. Analytical Technology and PAT. *BioProcess International* 5(1 Suppl):32-41.

Appendix A

Workshop Agenda

JULY 30, 2018

8:30 a.m.	**Registration**
8:55 a.m.	**Welcome** Elizabeth Boyle, The National Academies of Sciences, Engineering, and Medicine
9:00 a.m.	**Promise of Continuous Manufacturing** Kelvin Lee, University of Delaware
9:40 a.m.	**Continuous Manufacturing: Now and What the Future Might Hold at Biomedical Advanced Research and Development (BARDA)** Rick Bright, BARDA
10:20 a.m.	**Break**
10:30 a.m.	**Session 1: Business Case for Continuous Manufacturing** Art Hewig, Amgen Mauricio Futran, Janssen Franqui Jimenez, Sanofi **Panel Discussion: All Session Speakers** Moderated by: Gintaras V. Reklaitis, Chair, Purdue University
12:00 p.m.	**Lunch**
1:00 p.m.	**Session 2: Upstream Processing** Daisie Ogawa, Boehringer Ingelheim Shawn Barrett, Sanofi Eva Gefroh, Just Biotherapeutics **Panel Discussion: All Session Speakers** Moderated by: Ken Lee, MedImmune

2:30 p.m.	**Break**
2:45 p.m.	**Session 3: Downstream Processing** Lindsay Arnold, MedImmune Andrew Zydney, The Pennsylvania State University Mark Brower, Merck
	Panel Discussion: All Session Speakers Moderated by: Eva Gefroh, Just Biotherapuetics
4:15 p.m.	**Breakout Sessions** Business Case Upstream Processing Downstream Processing
5:15 p.m.	**Adjourn**

JULY 31, 2018

8:30 a.m.	**Goals for the Workshop from the U.S. Food and Drug Administration (FDA)** Janet Woodcock, FDA
9:10 a.m.	**Session 4: Product Manufacturing** Satoshi Ohtake, Pfizer Inc. William Whitford, GE Healthcare Tarit Mukhopadhyay, University College London
	Panel Discussion: All Session Speakers Moderated by: Michael Ladisch, Purdue University
10:40 a.m.	**Break**
10:50 a.m.	**Session 5: Integration** Chris Love, Massachusetts Institute of Technology (MIT) Richard Braatz, MIT Doug Richardson, Merck
	Panel Discussion: All Session Speakers Moderated by: Charles Cooney, MIT
12:20 p.m.	**Lunch**

1:00 p.m.	**Session 6: Regulatory and Quality Aspects of Continuous Manufacturing** Moheb Nasr, Consultant Lisa Connell-Crowley, Just Biotherapeutics Veena Warikoo, Roche
	Panel Discussion: All Session Speakers Moderated by: Keith Roper, University of Arkansas
2:30 p.m.	**Break**
2:40 p.m.	**Breakout Sessions** Product Manufacturing Integration Regulatory and Quality Aspects of Continuous Manufacturing
3:40 p.m.	**Report Back from Breakouts and Moderated Audience Discussion** Moderated by: Gintaras V. Reklaitis, Chair, Purdue University
5:00 p.m.	**Adjourn**

Appendix B

Biographies of Planning Committee Members, Speakers, and National Academies of Sciences, Engineering, and Medicine Staff

PLANNING COMMITTEE MEMBERS

Gintaras V. Reklaitis (NAE, *Chair*) is the Gedge Distinguished Professor of Chemical Engineering and courtesy professor of Industrial and Physical Pharmacy at Purdue University. His recent research interests include continuous manufacturing of pharmaceutical substances and products, including process monitoring, control, and modeling. Areas of long-term research are systems engineering applications to support batch and continuous operation and methodologies for plant- and enterprise-wide planning and optimization. His awards include the Warren K. Lewis Award of the American Institute of Chemical Engineers (AIChE) and the Pruitt Award of the Council of Chemical Research He has served on the AICHE Board of Directors and as Editor-in-Chief of *Computers & Chemical Engineering*. He is a member of the National Academy of Engineering and received his PhD from Stanford University.

Liuquan Chang is a group leader in the Global Regulatory Chemistry Manufacturing Controls (CMC) biologics at Merck, where she is responsible for providing regulatory leadership, oversight, and strategy to a team of Regulatory CMC Professionals in developing global Regulatory CMC strategy and executing submissions for biologics. She has more than 15 years of industry experience in CMC development, manufacturing, and regulatory submission of biological products in multiple global pharmaceutical companies. She had previously worked as the global regulatory lead for multiple investigational and marketed products at Sanofi regulatory CMC biologics. She received her PhD in Pharmaceutical Sciences from the University of Connecticut.

Charles Cooney is a Professor Emeritus at the Massachusetts Institute of Technology (MIT). His research and teaching span many aspects of biochemical engineering and pharmaceutical manufacturing. He holds more than 30 patents, has published more than 250 research papers, and has co-authored or edited 5 books. His honors include the 1989 Gold Medal of the Institute of Biotechnological Studies (London); the Food, Pharmaceutical, and Bioengineering Award from the American Institute of Chemical Engineers; the James Van Lanen Distinguished Service Award from the American Chemical Society's Division of Microbial and Biochemical Technology; and election to the American Institute of Medical and Biochemical Engineers and the Fellows of the American Chemical Society. He received his PhD in Biochemical Engineering from MIT.

Eva Gefroh is a Principal Scientist at Just Biotherapeutics, an integrated design company with a mission to improve global access to biotherapeutics by driving low-cost process development and manufacturing. She has almost 20 years of bioprocess development experience, including 15 years at Amgen in the Purification Process Development group. She was a group leader at Amgen, managing a filtration process development team for early and late stage molecules, and a commercial process development team leader. She was also a key member of the team that developed and implemented innovative technologies for a Manufacturing of the Future initiative with a focus on connected downstream processing, which included developing process control strategies for connected unit operations and devising viral clearance strategies to support validation of connected processes. In her current position, she continues to drive low-cost process development and manufacturing platforms through the assessment of new technologies, plant and equipment design, and process and economic modeling. She received her BS in Chemical Engineering from the University of Minnesota.

Ken Lee is an early and late-stage bioreactor scientist and Chemistry Manufacturing Controls functional lead at MedImmune. He joined MedImmune UK in 2011, where he introduced the site to the ambr15 technology and later perfusion technology in 2014. In 2015, Dr. Lee relocated to MedImmune US where he continued to work on continuous cell culture and integrated processing. He received his PhD in Biochemical Engineering from the University of Birmingham.

Keith Roper is the Charles W. Oxford Professor of Emerging Technologies in the Ralph E. Martin Department of Chemical Engineering at the University of Arkansas. His research examines advanced functionalities offered by active electrodynamics to next-generation nano-, bio-, and meta-materials, as well as frameworks for innovation such as nanoplasmonics and nanolithography, graphene, and van der Waals 2D materials, and functional membranes for biofuels and fuel cells. He is a fellow of the American Institute for Medical and Biological Engineering and a member of the Arkansas Academy of Science. He is an Associate Editor for Institute of Electrical and Electronics Engineers Transactions in Nanotechnology. He received his PhD in Chemical Engineering from the University of Wisconsin.

SPEAKERS

Lindsay Arnold is a downstream process scientist in the BioProcess Engineering Department at MedImmune. The primary focus of her group is technology scale up, new technology evaluation, and continuous processing development. Dr. Arnold received her doctorate from the Chemical and Biomolecular Engineering Department at the Georgia Institute of Technology.

Shawn Barrett is the Associate Scientific Director of United States Pharmacopeia in the Continuous Manufacturing Skill Center at Sanofi in Framingham, MA. He has more than 20 years of experience at various organizations in the biopharmaceutical and biotech industry, including Life Technologies BioProduction, Centocor Johnson and Johnson Research and Development, Eli Lilly, and the National Research Council Canada Biotechnology Research Institute. His bioprocess expertise includes fed batch and perfusion cell culture process development and scale up, cell culture medium development, bioreactor and facility design, and process control. He currently leads a team of scientists and engineers tasked with the optimization and intensification of a perfusion platform process for the enablement of integrated continuous manufacturing of protein therapeutics. He received his Bachelors of Applied Science in Chemical Engineering from the University of Waterloo in 1994.

Richard D. Braatz is the Edwin R. Gilliland Professor of Chemical Engineering at the Massachusetts Institute of Technology (MIT) where he does research in the design and control of continuous (bio) pharmaceutical manufacturing systems. He leads the Quality-by-Design and control systems activities in many (bio) pharmaceutical manufacturing efforts at MIT. He has consulted or collaborated with more than 20 companies, including Novartis, Pfizer Inc., Merck, Biogen, and Amgen.

Rick A. Bright is the Deputy Assistant Secretary for Preparedness and Response and the Director of the Biomedical Advanced Research and Development Authority (BARDA), which is a component of the Office of the Assistant Secretary for Preparedness and Response in the U.S. Department of Health and Human Services. He oversees the advanced development and procurement of medical countermeasures against an array of threats to national security and the public's health, including chemical, biological, and radiological threats, nuclear threats and pandemic influenza, and emerging infectious diseases. He began his career in vaccine and therapeutics development at the Centers for Disease Control and Prevention, with a focus on influenza viruses, antiviral drugs, and the development of novel assays for high throughput surveillance for resistance to antiviral drugs. For this work, he was a recipient of the Charles C. Shepard Science Award for Scientific Excellence. He has extensive experience in the biotechnology industry, where he served in senior leadership and executive management roles. He has also held senior scientific leadership positions in nongovernmental organizations where he championed innovative vaccine development and international vaccine manufacturing capacity expansion in developing countries. He serves as an international subject matter expert in vaccine, drug, and diagnostics development and has served as an advisor to the World Health Organization and the U.S. Department of Defense. He joined BARDA in 2010, and prior to becoming Director in late 2016, he served as Director of BARDA's Influenza and Emerging Infectious Diseases Division. He received his PhD in Immunology and Virology from Emory University and his BS in Biology and Physical Sciences from Auburn University.

Mark Brower is a Principal Scientist at Merck. His specialties include protein purification, process development, downstream processing, protein chromatography, expanded bed adsorption, simulated moving bed chromatography, and protein chromatographic modeling. He received his PhD in Chemical Engineering and Biotechnology from the University of Cambridge.

Lisa Connell-Crowley is Director of Downstream Process Design at Just Biotherapeutics, leading innovative groups of scientists and engineers to drive down the cost of biotherapeutics through streamlined, cost-effective process and plant design. She has 15 years of downstream process development experience, including 12 years at Amgen as a downstream group leader and Chemistry Manufacturing Controls process team leader for early and late stage antibodies and Fc fusion molecules. She is also a recognized expert in developing viral clearance strategies for monoclonal antibody processes, including understanding key parameters for retroviral clearance by low pH inactivation and various chromatography operations and designing strategies to assess viral clearance of continuous manufacturing processes. She received her PhD in Biochemistry from the Baylor College of Medicine.

Mauricio Futran (NAE) is the Vice President of Advanced Technology in the Global Tech Services group of Janssen Supply Chain at Johnson & Johnson, focusing on manufacturing process understanding and reliability. This is done by incorporating predictive modeling, in line measurements, data analytics, and other technologies into the full range of activities from research and development (R&D) through scale up, tech transfer, and life cycle management. The ultimate goal is model predictive control and Real-Time Release. Before joining Johnson & Johnson, he was Professor and Chair of Chemical and Biochemical Engineering at Rutgers University after working for 28 years in various positions in pharmaceutical product and process development at Merck and Co. and Bristol-Myers Squibb, where he was Vice President of Process R&D. His areas of expertise include all aspects of process development, technology transfer, validation, regulatory compliance, new product registration, external manufacturing, and partnership development. He is a member of the National Academy of Engineering, where he has been chair of its Chemical Engineering section, and has served on its peer committee, the Board of Chemical Sciences and Technology, and a National Research Council panel. As an American Institute of Chemical Engineers member, he has served on the awards committee. He has been a member and chair of the Princeton Chemical and Biological Engineering external board, and has been a member of the external boards for the University of Illinois at Urbana-Champaign, the Georgia Institute of Technology, and Rutgers University. He has Chemical Engineering degrees from Rice University and Princeton University.

Art Hewig is a Director of Process Development at Amgen, where he leads an integrated group that is designed to deliver end-to-end drug substance processes that robustly transfer to Amgen's commercial network. Prior to the role, he was heading Purification Process Development at Amgen. He has been with Amgen since 2002, where his initial focus was on developing purification processes for early and late stage molecules. During this time, he developed and implemented real-time process analytic technologies utilizing light scattering and connected downstream processing. He holds a PhD in Chemical Engineering from Carnegie Mellon University.

Franqui Jimenez is a Senior Director and head of Manufacturing and Technology at Sanofi. He is a senior executive with accomplished leadership, business, and people development skills in commercial biologics including process development and commercial technology transfer, startup, implementation, and production support. His work focuses on commercial biologics production, technical support, process development, mammalian-host biotechnology processes, monoclonal antibody production, and bacterial fermentation. He received his PhD in Chemical Engineering from the University of Wisconsin–Madison.

Michael R. Ladisch (NAE) is Director of the Laboratory of Renewable Resources Engineering, and Distinguished Professor of Agricultural and Biological Engineering with a joint appointment in the Weldon School of Biomedical Engineering. His BS (1973) from Drexel University and MS (1974) and PhD (1977) from Purdue University are in Chemical Engineering. His research applies principles of biological and chemical engineering to the study of proteins at interfaces for purposes of their fractionation, purification, and characterization using bioseparation techniques. This research addresses the manner in which proteins interact—with solid interfaces, with each other, or with molecules or microorganisms—principally in aqueous environments. The proteins studied include both enzymes and therapeutic molecules. A second and related thrust is development of bioprocesses by which renewable resources, broadly defined, may be transformed to low-carbon footprint bioproducts. The foundational science and engineering has been translated into industrial practice, numerous publications, 2 textbooks, and 20 patents that have been licensed to various companies. He is a member of the National Academy of Engineering and a Fellow of the American Institute for Medical and Biological Engineering, American Chemical

Society, American Association for the Advancement of Science, American Institute of Chemical Engineers, and the National Academy of Inventors.

Kelvin Lee is the Director of the Manufacturing USA National Institute for Innovation in Manufacturing Biopharmaceuticals and he is the Gore Professor of Chemical and Biomolecular Engineering at the University of Delaware. He previously served as the Director of the Delaware Biotechnology Institute. He received a BSE in Chemical Engineering from Princeton University and a PhD in Chemical Engineering from the California Institute of Technology (Caltech). He spent several years in the Biotechnology Institute at the Eidgenössische Technische Hochschule in Zurich, Switzerland, and also completed a postdoc in Caltech's Biology Division. Prior to his current appointment, he was on the faculty at Cornell University where he held the titles of Samuel C. and Nancy M. Fleming Chair Professor, Professor in the School of Chemical and Biomolecular Engineering, Director of the Cornell Institute for Biotechnology, and Director of the New York State Center for Life Science Enterprise.

Chris Love is a Professor of Chemical Engineering and a member of the Koch Institute for Integrative Cancer Research at the Massachusetts Institute of Technology (MIT). He is also an Associate Member of the Broad Institute, and an Associate Member at the Ragon Institute of Massachusetts General Hospital, MIT, and Harvard University. Dr. Love was a W.M. Keck Distinguished Young Scholar for Medical Research and a Dana Scholar for Human Immunology in 2009, a Life Sciences Research Foundation Postdoc Fellow (Gilead Sciences) in 2004, and a National Defense Science and Engineering Graduate Fellow from 1999 to 2002. He was also awarded the Foresight Distinguished Student Award in Nanotechnology in 2000, and is a Camille Dreyfus Teacher-Scholar. His team combines principles from chemical engineering, interfacial chemistry, and materials science to develop new micro- and nanotechnologies for addressing biological questions in immunology, microbiology, systems biology, and bioprocess engineering. One broad research objective is to improve the design and implementation of quantitative bioanalytical processes in order to maximize the knowledge gained about the heterogeneities and dynamics of individual cells within a complex population. The researchers employ these processes to characterize the pathogenesis and immunomodulation of complex immune responses in chronic human diseases, such as HIV/AIDS, multiple sclerosis, type 1 diabetes, food allergies, and cancer. A second objective is to develop new strategies for manufacturing biologic drugs to improve global access for patients. He received his PhD in Physical Chemistry from Harvard University.

Tarit Mukhopadhyay is a Senior Lecturer in the Department of Biochemical Engineering at University College London (UCL). He conducted his engineering doctorate in a joint venture between UCL and the Health Protection Agency, working on two vaccines of commercial interest, a novel Meningitis B vaccine based upon the outer membrane proteins of N. lactamica and the UK licensed

Anthrax vaccine. As part of his doctoral studies in a joint project with the Health Protection Agency and Professor Gary Lye he focused on the development of a microwell platform for rapid vaccine development. This platform was applied to the development of a new Meningitis B vaccine that is now in clinical trials. The platform's usefulness and robustness was also tested with the UK Anthrax Vaccine. He is currently working on a Japanese Encephalitis virus vaccine, hepatitis B, and the use of lentiviruses for gene therapy, with an interest in upstream and downstream processing and process analytical technology. He received his EngD in Biochemical Engineering from UCL.

Moheb Nasr is the Principal of his own consulting company, Nasr Pharma Regulatory Consulting (NPRC), which was established in February 2018. NPRC provides strategic regulatory consulting with a focus on organizational aspects, manufacturing, pharmaceutical development innovation, and Chemistry Manufacturing Controls (CMC) regulatory transformation. He continues to play a leading role in global regulatory harmonization and introduction of modern pharmaceutical manufacturing platforms, including continuous manufacturing. He also represents Pharmaceutical Research and Manufacturers of America at The International Conference on Harmonisation of Technical Requirements for Registration of Pharmaceuticals for Human Use (ICH), serving as an ICH Expert Working Group Topic Lead. He retired from GlaxoSmithKline (GSK) in August 2017. At GSK, he was responsible for the development and the execution of GSK CMC regulatory strategy and served as a member of GSK leadership and governance boards accountable for product development, manufacturing and supply, quality, and regulatory oversight. In recognition of his scientific contributions and visionary leadership, he is an elected GSK Senior Research Fellow. Prior to joining GSK, he spent more than 22 years at the U.S. Food and Drug Administration (FDA). At FDA, he served as a bench chemist, regulatory scientist, director of FDA's analytical laboratories, and FDA/Center for Drug Evaluation and Research (CDER) CMC regulatory office head. He also established and led FDA's Office of New Drug Quality Assessment. He received his Pharmacy degree at the University of Cairo, Egypt, and his PhD in Chemistry at the University of Minnesota.

Daisie Ogawa is the head of the Upstream continuous manufacturing team at Boehringer Ingelheim, working under Jon Coffman. The group focuses on perfusion cell culture process development and scale up, with recent efforts emphasizing process intensification. The process is developed in a proprietary manufacturing platform called the iSKID™ system; this work is performed in collaboration with Pfizer.

Satoshi Ohtake is the Senior Director of Pharmaceutical Research and Development at Pfizer, where he oversees all biotherapeutic drug product development activities. Prior to joining Pfizer, he held positions of increasing responsibility at Nektar Therapeutics and Aridis Pharmaceuticals, specializing in the development of novel processing technologies and formulation of various modalities ranging from small molecules to live attenuated vaccines. He also serves as an executive board member of a nonprofit organization, adjunct faculty member at the University of Kansas, and Editorial Advisory Board member of peer-reviewed journals. He received his BS from the California Institute of Technology and PhD from University of Wisconsin–Madison, both in Chemical Engineering.

Douglas Richardson is a Principal Scientist in Bioprocess Development at Merck Research Labs in Kenilworth, New Jersey. His current research focuses on the evaluation and implementation of new automation and analytical technologies for the characterization of protein therapeutics. Recently, this research has focused on expanding the large molecule Process Analytical Technology (PAT) toolkit to support current and future biotherapeutic manufacturing platforms. Dr. Richardson received his PhD in Analytical Chemistry from the University of Cincinnati in the lab of Dr. Joseph A. Caruso.

Veena Warikoo is a Senior Biotherapeutics Chemical Manufacturing Controls (CMC) Development leader at Roche. She has 20 years of experience in both strategic planning and execution for big pharma and startups in early and late phase process development, manufacturing, regulatory, and pharmacokinetics. Her areas of expertise include early and late phase bioprocess development, integrated continuous biomanufacturing, pharmacology/safety/toxicokinetics, and application of computational models, scale up and scale down of bioprocesses, fill-finish, cgmp, manufacturing technical support, and organizational development. She received her PhD in Microbiology and Biochemistry from the University of Oklahoma.

William Whitford is the Strategic Solutions Leader of GE Healthcare in Logan, Utah, with more than 20 years of experience in biotechnology product and process development. He joined the company as an R&D Leader developing products supporting protein biological and vaccine production in mammalian and invertebrate cell lines. Products he has commercialized include defined hybridoma and perfusion cell culture media, fed-batch supplements, and aqueous lipid dispersions. An invited lecturer at international conferences, he has published more than 300 articles, book chapters, and patents in the bioproduction arena. He now enjoys such activities as serving on the editorial advisory board for BioProcess International.

Janet Woodcock is the Director of the Center for Drug Evaluation and Research at the U.S. Food and Drug Administration (FDA), making sure that safe and effective drugs are available to improve the health of people in the United States. Her center evaluates prescription and over-the-counter drugs before they can be sold and oversees their testing in clinical trials, provides health care professionals and patients with the information that they need to use medicines wisely, ensures that drugs, both brand-name and generic, work correctly and that their health benefits outweigh their known risks, and takes action against unapproved, contaminated, or fraudulent drugs that are marketed illegally. Since joining FDA in 1986, she has led many of FDA's drug initiatives and introduced the concept of risk management in 2000 as a new approach to drug safety. Since 2002, she has led the Pharmaceutical Quality for the 21st Century Initiative, FDA's highly successful effort to modernize drug manufacturing and its regulation. In 2004, she introduced FDA's "Critical Path" Initiative, which is designed to move medical discoveries from the laboratory to consumers more efficiently. She received her medical degree from Northwestern University Medical School, and her undergraduate degree from Bucknell University. She has held teaching appointments at The Pennsylvania State University and the University of California, San Francisco.

Andrew Zydney is the Distinguished Professor of Chemical Engineering and the Director of the Center of Excellence in Industrial Biotechnology at The Pennsylvania State University. He served as Head of the Department of Chemical Engineering from 2004-2014. He received his BS in Chemical Engineering from Yale University in 1980 and his PhD from the Massachusetts Institute of Technology in 1985. He was also a faculty member in the Chemical Engineering Department at the University of Delaware from 1985-2001. His research is focused on membrane science and technology, with a particular emphasis on applications of membranes in bioprocessing. He is the most recent recipient of the Gerhold Award for Excellence in Separation Science from the American Institute of Chemical Engineers (AIChE) and is a past recipient of the Excellence in Biological Engineering Publications Award from AIChE. He is the Editor-in-Chief of the *Journal of Membrane Science*, and he serves on the Editorial Boards for *Separation and Purification Reviews*, *Separation Science and Technology*, *Current Opinion in Chemical Engineering*, and *Biotechnology and Bioengineering.*

NATIONAL ACADEMIES STAFF

Elizabeth Boyle has more than 15 years of experience in environmental health and epidemiology. She currently works as a Program Officer on the Board on Environmental Studies and Toxicology within the National Academies' Division on Earth and Life Studies. Formerly, she was an Environmental Health Scientist at Westat, where she supported the Environmental Protection Agency, the National Institute of Child Health and Development, and the National Cancer Institute by completing other environmental epidemiology related projects. Prior to her tenure at Westat, she was a student epidemiologist at the Minnesota Department of Health and an Industrial Hygienist at a consulting firm in Cincinnati. She serves as Chair of the Nominations Committee for the International Society of Exposure Science. She is also a fellow of the Bloomberg American Health Initiative at the Johns Hopkins Bloomberg School of Public Health, where she is pursuing a doctor of public health in environmental health. She has an MPH in Environmental Epidemiology, a BS in Biology, and she is a Certified Industrial Hygienist.

Marilee Shelton-Davenport is a seasoned science policy program director who joined the Board on Chemical Sciences and Technology at the National Academies of Sciences, Engineering, and Medicine in 2016 after 17 years with the National Academies' Board on Life Sciences. She has an extensive portfolio of work at the intersection of chemistry, biology, and health including impactful activities that serve a variety of federal agencies. A natural at building bridges and working across traditional siloes, her work includes Consensus Study Reports and the development of innovative activities and communication approaches to engage a wide breadth of scientists and stakeholders in impactful dialogues related to emerging technologies, environmental health and toxicology, and chemical and biological defense. She received her PhD in Pharmacology from the University of North Carolina at Chapel Hill and her BS in Biochemistry from Clemson University.

Jarrett Nguyen is a Program Specialist in the Office of the President at the National Academy of Medicine. Formerly he was a Senior Program Assistant for the Board on Chemical Sciences and Technology. Mr. Nguyen began working for the National Academies of Sciences, Engineering, and Medicine in October 2016. He graduated from James Madison University in May 2015 with a BS in Geology and Environmental Science with a minor in Geographic Science.

Appendix C

Running Themes Across Workshop Sessions

TABLE C-1 Technologies and Issues Raised with Respect to Regulations

Topic	Organization	Speaker	Technologies/Issues Addressed
Science of Pharma Manufacturing	FDA	Janet Woodcock	FDA's interest in furthering science of continuous manufacturing requires collaboration with industry, academia, and other federal sponsors.
Medical Countermeasures	BARDA	Rick Bright	BARDA focuses on rapidly producing medical countermeasures to changing natural and manmade security threats.
United States and International Regulatory Considerations	Nasr Regulatory Consulting	Moheb Nasr	White papers are available on both small molecules and biologics. There is a regulatory framework adequate to support continuous manufacturing: Good Manufacturing Practices (GMPs), the critical quality attributes (CQAs), and the Points to the International Conference on Harmonisation of Technical Requirements for Registration of Pharmaceuticals for Human Use (ICH) Q13 guideline.
Real-time Product Release and Adaptive Process Control; Process Analytical Technology	Merck	Doug Richardson	Merck's experience: measure CQA; model Process; apply appropriate control strategy; use lessons learned to optimize process; and remember that there is "no path to real-time release without process analytical."
Viral Safety Guideline ICH Q5A	Just Biotherapeutics	Lisa Connelly-Crowley	Robust and effective viral clearance of retrovirus-like particles is critical.

TABLE C-2 Technologies and Issues Raised with Respect to Regulations

Topic	Organization	Speaker	Technologies/Issues Addressed
Perfusion	Amgen	Art Hewig	Shrinking the manufacturing footprint and capital investment is needed. Perfusion gives high titers; 25 to 60 g/L continuous manufacturing will be company specific.
One-room production, critical quality attributes	Janssen	Maricio Futran	One room operation enables continuous monitoring; ability to monitor CQA during (rather than after) production is critical.
Perfusion	Sanofi	Shawn Barrett	Sanofi is moving from micro-carrier Chinese hamster ovary (CHO) culture to suspension CHO culture with continuous capture. Intensified perfusion platform yields 100 × increase in productivity.
Continuous downstream processing with continuous upstream processing demonstrated	MedImmune	Lindsay Arnold	MedImmune has operated continuous downstream processing for two weeks: Protein A capture, low pH virus inactivation, filter train, multi-column chromatography, single pass buffer exchange, and final concentration has been run with continuous upstream.
Developed integrated, skid mounted single-use technology	Boehringer Ingelheim	Daisie Ogawa	The process has run continuously with no "air-gaps;" process has increased productivity by 2 to 6 × 4 × reduction in media volume, with potential to manufacture multiple products.
Using experiences from food industry	Pfizer	Satashi Ontake	Note pros and cons of integrated versus piecemeal continuous processing.

TABLE C-3 Unit Operations and Input to Process Models

Topic	Organization	Speaker	Technologies/Issues Addressed
Science of Pharmaceutical Manufacturing	FDA	Janet Woodcock	A challenge is the integration of downstream unit operations in an effective manner that satisfies purity requirements.
Making the business case for continuous manufacturing	Janssen Pharmaceuticals	Mauricio Futran	Known unit operations lower the hurdle to implementing continuous processes.

(Continued)

TABLE C-3 Continued

Topic	Organization	Speaker	Technologies/Issues Addressed
Downstream processing	Merck	Mark Brower	Labs can be used to test unit operations and help provide information to design new systems.
Achieving full integration	Massachusetts Institute of Technology	Richard Braatz	Achieving full automation requires complete understanding of each unit operation. In silico models and dynamic models can increase understanding of unit operations for plant-wide simulation and control design.
Considerations for integration	Roche	Veena Warikoo	Intensified processes means fewer unit operations, but there needs to be more columns and column cycles and liquid flow rates could be controlled continuously so that they are synchronous as a means of mitigating deviations.

TABLE C-4 Discussion of Business Case in Technical Sessions

Topic	Organization	Speaker	Technologies/Issues Addressed
Downstream processing	MedImmune	Lindsay Arnold	The business cases vary based on the particular set of drivers that are important to the company, such as safety, cash flow, capital reduction, space constraints, and capital improvement, and will vary between clinical and commercial settings.
Using digital biomanufacturing to support product manufacturing	GE Healthcare	William Whitford	Digital biomanufacturing should also support business continuity with incident control, management, and reporting capabilities, and in-line or on-line, real-time, orthogonal process monitoring and adaptive control.
Considerations for integration	Roche	Veena Warikoo	There will not be a business case for continuous manufacturing without process intensification. Product intensification challenges business models, opening opportunities for new patentable products, process chemistry, and change to just in time or distributed manufacturing.